94' ✓

WILD
INDONESIA

WILD INDONESIA

The wildlife and scenery of the Indonesian archipelago

Photographs by **GERALD CUBITT**
Text by **TONY and JANE WHITTEN**

WWF

Produced in association with the World Wide Fund for Nature

The MIT Press
Cambridge, Massachusetts

First MIT Press edition, 1992
© 1992 New Holland (Publishers) Ltd
© 1992 in photographs Gerald Cubitt (except individual photographers as credited on page 8)
© 1992 in text Anthony and Jane Whitten

Library of Congress Cataloging-in-Publication Data

Cubitt, Gerald S.
     Wild Indonesia: the wildlife and scenery of the Indonesian archipelago/
photographs by Gerald Cubitt; text by Tony and Jane Whitten. — 1st MIT Press ed.
     p. cm.
     Includes bibliographical references and index.
     ISBN 0-262-23165-4
     1. Natural history—Indonesia. 2. National parks and reserves—Indonesia. 3. Ecology—
Indonesia. 4. Indonesia—Description and travel. I. Whitten, Tony, 1953-   . II. Whitten, Jane.
III. Title.
     QH186.C83 1992   92-14891
     508.598—dc20   CIP

Commissioning Editor: Charlotte Parry-Crooke
Project Manager: Ann Baggaley
Editorial Assistants: Charlotte Fox, Leigh Williams
Designer: Behram Kapadia
Cartography: Julian Baker, Stephen Capsey, Gillian Ballington: Maltings
Partnership
Index: Janet Dudley

Typeset by Bookworm Typesetting, Manchester
Reproduction by Scantrans Pte Ltd, Singapore
Printed and bound in Singapore by Kyodo Printing Co (Singapore) Pte Ltd

CONTENTS

FOCUS ON SUMATRA *page 80*

FOCUS ON THE MOLUCCAS *page 166*

FOCUS ON THE LESSER SUNDAS *page 174*

FOCUS ON IRIAN JAYA *page 182*

Photographic Acknowledgements

The publishers and photographer extend their thanks to the following people who kindly loaned their photographs for inclusion in this book. All the photographs in the book, with the exception of those listed below, were taken by Gerald Cubitt.

Lynn Clayton: page 161 (bottom left).

Robert Harding Picture Library (James Green): page 146.

Andrew Laurie: page 18 (below right).

Linda Pitkin: page 62 (below); page 64 (all three subjects); page 65 (above); page 157 (all five subjects); page 158 (all three subjects); page 159 (all four subjects).

H.D. Rijksen: page 18 (below left); page 30 (right); page 41; page 49 (right); page 93 (top).

Tony Whitten: page 29 (left); page 33; page 40 (above right and below); page 86 (below); page 87 (centre left and right); page 90 (below right); page 127 (above left and right); page 147 (above); page 148 (above left and below); page 149 (all six subjects); page 164 (centre); page 165 (bottom right).

WWF Photolibrary (Dr P.K. Anderson): page 60 (below). *(Eugen Schuhmacher):* page 133 (below left).

Illustrations appearing in the preliminary pages are as follows:

HALF-TITLE: Young Agile Wallaby (*Macropus agilis*) from southern Irian Jaya.
FRONTISPIECE: Sunset over the Togian Islands, Central Sulawesi.
PAGE 5: View from Samosir Island in Lake Toba, North Sumatra.
PAGE 6: Wild ginger (*Riedelia* sp.), endemic to New Guinea and the Moluccas.
PAGE 7: Australian Pelicans (*Pelecanus conspicillatus*) from Irian Jaya.

Acknowledgements

Interest in this project has been forthcoming from numerous sources around the world. The publishers would especially like to thank the sponsors, contributors and consultants for their involvement. The authors, photographer and publishers would like to express their gratitude to the following for their generous and valuable assistance during the preparation of this book:

Joop Avé, Director-General of Tourism, Jakarta
World Wide Fund for Nature (WWF)
Mandarin Oriental, Jakarta

INDONESIA
Drs Effendy Sumardja, Head of Forestry, Bali
Udin Saifuddin, Marketing Director, Directorate General of Tourism, Jakarta
Peter Pangaribuan and Zain Sumedy, Directorate General of Tourism, Jakarta
The Head of Tourism, Irian Jaya
The Central and Field Staff of the Offices of Natural Resources and Nature Conservation throughout Indonesia
Garuda Indonesia
Merpati Nusantara Airlines
Indoavia Air Charter
Dr Kathy MacKinnon and Dr Charles Santiapillai, WWF Indonesia
Drh Linus Simanjuntak, Ragunan Zoo, Jakarta
Dr Biruté Galdikas, Orangutan Research and Conservation Project, Tanjung Puting
Dr Kate Monk, EMDI Project, Lombok
Max Zieren, Asian Wetland Bureau, Bogor
Putri Hidayat and Ningsih Chandra, Mandarin Oriental, Jakarta
Aristedes Katoppo and Aco Manafe, MUTIARA (Surat Kabar Mingguan)
Don Hasman
Robby Semeru
Des Alwi
J.B. Ridgeway
Yakobus Karacap
Jack Daniels, Spice Island Cruises
Stanley Allison, P.T. Aerowisata
Drg Halim Indrakusuma, Pacto Tours
Heyder Souisa, Pacto Tours

David Heckman, Sobek Expeditions
Syamsuarni Syam and Treesnawarty Madaning, Insatra Exclusive Tours, Sulawesi
Freeport Indonesia
Hotels Nusa Dua Beach and Sanur Beach, Bali
Senggigi Beach Hotel, Lombok
Heritage Sarana Resorts

SOUTH-EAST ASIA AND AUSTRALASIA
Dr John MacKinnon, WWF Hong Kong
Dr Ronald Petocz, WWF Philippines
Dr Peter Ng, National University of Singapore
Dr Tim Flannery, Australian Museum, Sydney, Australia

EUROPE
Peter Jackson
Dr John Dransfield, Royal Botanic Gardens, Kew
Dr Mark Cheek, Royal Botanic Gardens, Kew
Dr Philip Cribb, Royal Botanic Gardens, Kew
Dr George Argent, Royal Botanic Garden, Edinburgh
Rosemary Smith, Royal Botanic Garden, Edinburgh
Maureen Warwick, Royal Botanic Garden, Edinburgh
Dr Roy Watling, Royal Botanic Garden, Edinburgh
Dr Colin McCarthy, Natural History Museum, London
Dr Dick Vane-Wright, Natural History Museum, London
Lynn Clayton, Oxford University
Dr Jim Comber
John Edwards Hill
Dr Jan van Tol, Natural History Museum, Leiden
Dr Tim Whitmore, Cambridge University

PREFACE

TOWARDS A SUSTAINABLE FUTURE

Indonesia is a country of enormous natural diversity, both of habitats and species. It has a total land area of 193 million hectares of which roughly three-quarters is covered in forest. With a growing population, now standing at 180 million inhabitants, and its status in the low income countries of the world, Indonesia needs to place a priority on development if it is to be released from its poverty trap.

While the population's standard of living must be raised it is essential that as much as possible of the country's natural environment is conserved. Indonesia's progress is rooted in its human resources and the riches of the land. It is therefore appropriate that we pursue a management strategy combining industrialization with a high regard for the sustainable use of natural resources and allowing precious wild regions to be safeguarded on the path towards a successful future.

Wild Indonesia serves as a record of many of the beautiful wilderness areas of the Indonesian archipelago and highlights some of the problems with which they are faced. I hope that this book will increase awareness of our natural heritage and encourage all of us to understand as fully as possible the issues affecting its preservation.

PROFESSOR DR. EMIL SALIM
MINISTER OF STATE FOR POPULATION AND THE ENVIRONMENT

FOREWORD

Indonesia is a land of extraordinary beauty. Located along one of the most active segments of the 'Ring of Fire', that immense volcanic circle around the Pacific Ocean, Indonesia is characterized by almost countless islands and a diversity of land forms which may not be surpassed anywhere in the world. From its coral reefs and shorelines to the top of the glacier-clad Carstenz Peak in Irian Jaya, Indonesia's rain forests, savannahs, mountains, seas and the lands in between are home to a diversity of life found in few other areas of the globe.

From Indonesia comes the famous Komodo Dragon; the metre-wide *Rafflesia*, the world's largest flower; the Orangutan, the 'man of the forest', whose Malay name has found its way into the English language; numerous strange marsupials; countless varieties of insects, including the largest variety of butterflies of any country; and a host of other natural wonders which are unique to Indonesia.

The scientists call this abundance 'biodiversity', and its protection, world wide, has become one of the major issues of the last decade of the twentieth century. Indonesia is one of the most important centres of biodiversity in the world. Although the country covers only 1.3 per cent of the Earth's land surface, its territory includes over 10 per cent of the world's flowering plant species, 12 per cent of the world's mammal species, 16 per cent of all reptile and amphibian species, 17 per cent of the world's bird species, and over 25 per cent of the planet's fish species.

Protection of such biodiversity is an essential part of WWF's world wide mission to achieve the conservation of nature and ecological processes, so that humans might be able to continue to live in harmony with nature on Planet Earth. Thus, we view Indonesia as a key country in the increasing global efforts to preserve genetic, species and ecosystem diversity.

For these reasons, *Wild Indonesia* is a welcome addition to the literature on this beautiful and important country. I am confident that its appearance at this time will stimulate greater knowledge of, and interest in, the beauty of nature, especially in Indonesia. WWF is committed to assist in the preservation of this beauty, and of the genetic heritage which it holds, for the future benefit of all. We welcome the growing concern of others in meeting this challenge.

CHARLES DE HAES
DIRECTOR GENERAL
WWF – WORLD WIDE FUND FOR NATURE

Introduction to Indonesia

Patterns of Islands

It is said that Indonesia comprises 13,667 islands. Even on a tourist guide map there are hundreds, and the larger the scale of the map the more idyllic little islands one discovers. On nautical charts, there are parts of Indonesia which are fairly littered with islands, many of them uninhabited and uninhabitable for want of fresh water and shelter. The country is dominated, however, by just five islands: Sumatra, Kalimantan (Indonesian Borneo), Java, Sulawesi and Irian Jaya (Indonesian New Guinea). These are the cultural, economic, development and political centres for this, the world's largest archipelago.

The country divides quite neatly into two parts. The first is Western Indonesia, or the Sunda area, consisting of Sumatra, Kalimantan, Java and Bali. These islands have many biological and cultural similarities and together with Peninsular Malaysia, the rest of Borneo, the Philippine island of Palawan, and the southern part of the Thailand Isthmus, form an area sometimes known as Sundaland. Sundaland lies on the Sunda Shelf, an extension of the Asian continent where the sea is shallow, 40 metres (130 feet) deep or less. These areas have been connected with one another now and again during the geological past as the sea level has risen and fallen, alternately allowing and preventing the dispersal of animals and plants from the Asian continent and between the islands. The second part, Eastern Indonesia, begins east of the Sunda Shelf, from Sulawesi in the west to Irian Jaya in the east, with the islands of the Moluccas and Lesser Sundas between them, none of which has ever been connected with a major land mass. Irian Jaya, on the other hand, sits on the Sahul Shelf, an extension of the northern part of Australia, with which it had land connections when the sea level was lower.

The difference in geological conditions between the two parts is reflected in their wildlife. This was first recognized by the thoughtful nineteenth-century British zoologist and inveterate traveller, Alfred Russel Wallace, whose researches here over 120 years ago gave the world its first understanding of this area's biological significance. When he crossed to Lombok in the Lesser Sundas from Bali in 1856 he wrote, 'I now saw for the first time many Australian forms that are quite absent from the islands westwards'. The dissection of Indonesia by what has become known as Wallace's Line between Bali and Lombok and between Borneo and Sulawesi used to be held as a fundamental and all-embracing concept in biogeography, but it is in fact only useful for describing the distribution of certain groups of larger animals. For other groups, their distributions are better explained by the relative proximities of the Philippines, Lesser Sundas and Moluccas.

Eastern Indonesia is made up of two distinct areas. One is Irian Jaya, the western part of New Guinea, and the other is the intermediate zone of small islands, named after Wallace and known as Wallacea, which lies between the Sunda and Sahul Shelves. Wallacea is neither entirely Asian nor entirely Australian, nor is it simply a mixture of the two but rather contains many creatures which are not found in either Asia or Australia.

Climate

Indonesia's climate is generally moist with abundant rainfall and high temperatures. Through the year the rainfall, wind and the amount of cloud cover will vary, but temperature changes little. Indonesia lies along the equator which results in a large area of relatively uniform conditions but, given the sheer size and diversity of the archipelago, it is not surprising that there is variation; some areas are very dry, and the high areas are relatively chilly.

The wettest places in Indonesia are around Mount Slamet in Central Java which receive more than 7,000 millimetres (275 inches) of rain annually, equivalent to the height of a two-storey building. The northern and southern slopes of the central highlands in Irian Jaya, the central mountains in the Bird's Head Peninsula in Irian Jaya to the east, and the slopes of Sumatra's main mountain range, receive 5,000–6,000 millimetres (200–240 inches). The sheer quantity of rain that can fall in a full-blown tropical storm is hard to imagine if you have not been caught out during one, but it is similar to standing under a tepid shower, switched on full. Little wonder that everything stops as people dive for cover. The worst storms on record in Indonesia resulted in 802 millimetres (31 inches) of rain falling in one day, and 3,220 millimetres (127 inches) falling over thirteen consecutive days with 80 millimetres (3 inches) falling in just thirty minutes.

At the other extreme, the Palu Valley in Central Sulawesi receives less than 500 millimetres (20 inches) of rain a year, and several locations in the Lesser Sundas are also termed arid, receiving less than 1,000 millimetres (40 inches), with many of those islands being dry, not receiving more than 2,000 millimetres (80 inches). Most of the country, however, is moist or wet, receiving an acceptable 2,000–4,000 millimetres (80–160 inches) of rain per year.

Temperatures are largely uniform through the year, not changing much from 30°–34°C (86°–93°F) at sea level. Likewise, relative humidity is uniformly high across most of the country, exceeding 80 per cent in most months, except in the drier islands in the south-east. In most of Western Indonesia and Irian Jaya, humidity approaches 100 per cent during the night and falls to 30–55 per cent at midday.

OPPOSITE PAGE The dark, cool green of the rain forest conveys the mystery of wild Indonesia. This view is taken in Mount Leuser National Park, Sumatra.

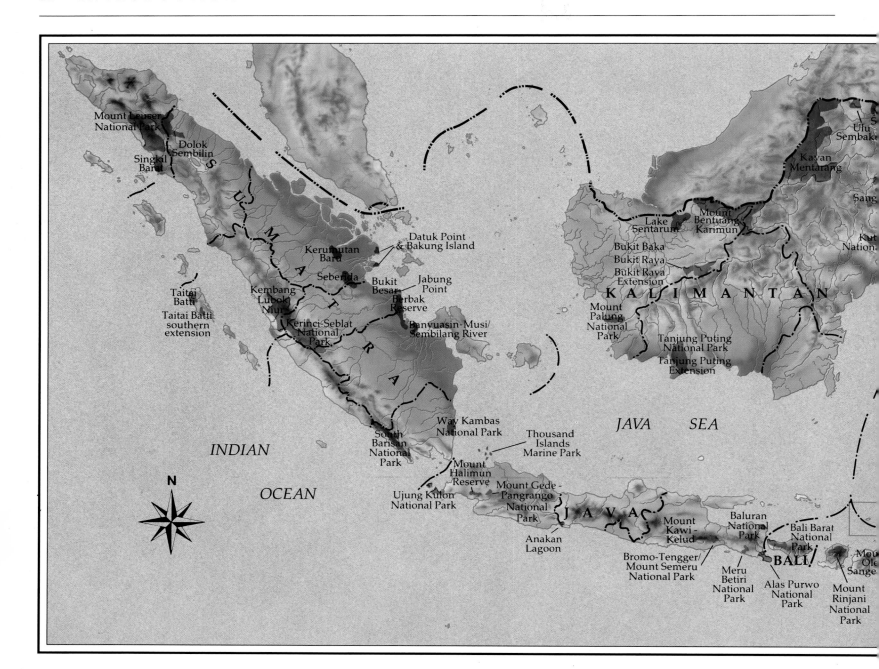

Diversity of Habitats

Indonesia is well known for its lowland forests and the varied wildlife they contain, but the archipelago straddles the equator for 5,000 kilometres (over 3,000 miles) and there is much more to it than just tropical rain forest. There is a great variety of physical conditions to which plants and animals have adapted in myriad ways.

The natural vegetation of most of Indonesia's sheltered coasts is mangrove swamps with tangled roots and thigh-deep mud. This is backed by other forms of swamp, standing on different depths of peat. On exposed coasts one encounters steeply shelving rocky shores, pounded by persistent and often ferocious waves. Inland there are numerous distinct types of forest determined by the soil and slope of the land, such as limestone forest, heath forest and ultrabasic forest. As altitude increases up a mountain so the trees get smaller and more gnarled and the plants show increasing similarities to those from temperate parts of the world. At the extreme of altitude in Irian Jaya, on the summit of Puncak Jaya at 5,039 metres (16,532 feet) the conditions are too cold for any permanent life. In Eastern Indonesia, there is also a range of dryland habitats which one

does not generally associate with Indonesia, such as savannah and deciduous forest.

In addition to these, there are habitats which bear the obvious mark of man, such as scrubby bush and grasslands, as well as the agricultural ecosystems of wet or dry ricefields, plantations, or mosaics of shifting agriculture. Indonesia's great diversity is also seen in the different patterns of land use among the islands. One of the most dramatic contrasts between the main islands is found in the percentage of land covered by forest. The two extremes are Irian Jaya with 84 per cent forest and less than 1 per cent intensively used land, and Java and Bali with less than 10 per cent forest and 73 per cent intensively used land.

Diversity of Life

Indonesia is enormously rich in both animal and plant life. Although it is an extensive country, it covers only just over 1 per cent of the world's land surface. Despite this, it is possible to find within its boundaries 10 per cent of the world's plant species, 12 per cent of the world's mammal species, 16 per cent of the world's reptile and amphibian species, and 17 per cent of the world's bird species.

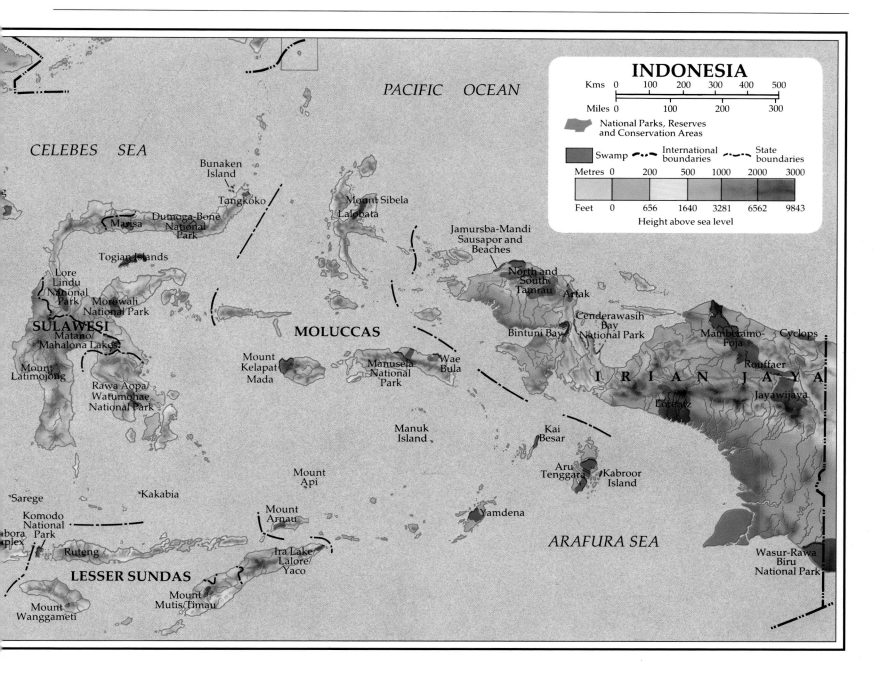

Not only are the percentages impressive, but many of the species concerned are endemic – that is, found here and nowhere else. This applies to about 430 of the roughly 1,500 bird species, 200 of the 500 mammal species, and over half of the 350 species of the economically important dipterocarp trees, with 155 species endemic to the island of Borneo alone. Although some countries have more species than Indonesia, it is unlikely that any has a more diverse and unique wildlife, because diversity measures not just quantity but variety. The spread of Indonesia across from Asia to Australia, and from torrid coastal swamps to glaciers, gives it an unmatched status among the countries of the world.

The species living today are only a snapshot of the life that has adorned the Earth through the millennia. About 700,000 years ago, when the Woolly Rhinoceros and Mammoth roamed northern Europe, the forests and plains of Java were inhabited at different times by eight species of elephants, as many as three living at one time. There were also three species of rhinoceros, three species of pig, large sabre-toothed cats, and a giant pangolin or scaly anteater. Try to picture also, in this unlikely community, groups of hippopotamuses and hyenas, animals which are now strictly African.

Caves in South Sulawesi have revealed something of the animals alive and sharing the country with man about 30,000 years ago. A number of these were giants, such as a huge pig and a tortoise with a shell nearly 2 metres (6½ feet) long, considerably larger than the giant tortoises found today in the Galapagos Islands. Elsewhere in the archipelago there is evidence of giant rats from Timor and Flores, and of a large lizard on Java and Timor, comparable in size to the fabled 'Dragon' living on Komodo Island today. Largest of all the land animals were the stegodonts, which looked much like modern elephants except that the males had large, curving tusks which grew so close together that the trunk must have been draped over one side. In contrast there was also a pygmy stegodont no taller than a man, remains of which have also been found on Flores and Timor. The different types of elephants found in Indonesia and on the neighbouring islands of the Philippines and Taiwan probably swam across quite wide expanses of water. (Modern elephants, too, are able to swim well, and the world record for distance is held by a luckless individual which was washed overboard from a ship in 1856, 48 kilometres (30 miles) from the South Carolina coast. Local residents must have been aghast when this exhausted beast walked out of the sea!

LEFT Only three places in the tropical world can boast glaciers and one of these is Irian Jaya, where an expanse of snow and ice is found near Indonesia's tallest peak, Puncak Jaya or Mount Carstenz, 5,039 metres (16,532 feet).

OPPOSITE PAGE, LEFT Volcanoes are common in many parts of Indonesia. Springs heated by volcanic activity, such as this one in the crater of Mount Rinjani on Lombok, attract visitors and the cooler pools are pleasant to sit in and are good for the skin.

BELOW Rice from irrigated paddy fields is the staple food of most Indonesians. Buffalo, as seen here near Bukittinggi in West Sumatra, are common draught animals.

Many cultivated plants in temperate regions originate from the mountains of Asia, such as this *Begonia* (TOP) and *Rhododendron zoelleri* (ABOVE) from Irian Jaya.

Much less is known about the vegetation of the past than the animals, but tree leaves and fruit from about 15 million years ago have been found. Climatic changes over the last million years have caused considerable changes in the vegetation although the only moderately reliable information we have is from the last 30,000 years or so. In that time both temperature and rainfall have varied. During the drier and more seasonal periods, there was much less rain forest than in recent times. The rain forest in its present form has not been in place from time immemorial.

Regrettably, Indonesia has already known the tragedy of modern extinctions. The Javan Wattled Lapwing (*Vanellus macropterus*) and the Caerulean Paradise Flycatcher (*Eutrichomyias rowleyi*) of the Sangihe Islands near Sulawesi appear to have gone forever, as have the small subspecies of tiger on Bali and almost certainly on Java. More extinctions are probably imminent. Some species may now be represented by populations which are quite incapable of sustaining themselves in Indonesia in the long term, such as the Javan Gibbon and the Sumatran Elephant, and so in some ways they can already be regarded as ecologically or pragmatically extinct. Similarly, there are two trees in the Bogor Botanical Gardens, just south of Jakarta, which represent the last living specimens of their species. The term of the 'living dead' has been applied to such species.

Diversity of Peoples

Perhaps one of the best known facts about Indonesia is the very high population density of the island of Java. The statistics from 1985 reveal that Java occupies just 6.9 per cent of the land area of Indonesia but has 60 per cent of the country's population, living at a density of 758 people per square kilometre. This is an average figure and hides even higher densities in towns and cities. Even so, there are large areas of rural and wild Java which have few people, and it should be remembered that the density of people in Holland, for example, is about 2,500 people per square kilometre. Sumatra's percentage of population and of land area are more or less equal at about 20 per cent, but the enormous expanses of both Kalimantan and Irian Jaya have very small populations living at an average of fifteen and four people per square kilometre respectively. Again, these average figures hide the differences between the more densely populated towns and the extremely thinly populated interiors.

These skewed proportions are not necessarily inappropriate. Java is blessed with wonderfully fertile soils and is able to support many more people than, say, the extensive swamps, highlands and steep mountains of Irian Jaya.

The motto of the Indonesian Republic is 'Unity in Diversity'. The unity relates to the way in which the country has held

ABOVE A Batak woman from near Lake Toba displaying her colourful *ulos* shawls which are worn and presented at ceremonial and other important occasions.

BELOW A young girl searching for nits in the hair of a member of her clan on the island of Siberut in the Mentawai group west of Sumatra. Groups of these isolated people have resisted undue change to their lifestyle but have taken from civilization those elements they feel are of benefit.

RIGHT Two Dayak women chatting at a trading post in the highlands of Borneo. Note the elongated ear lobes that would once have borne heavy earrings.

BELOW A Wana man in Morowali National Park, Sulawesi, enjoying the sweet honey of a wild bees' nest.

RIGHT Men collecting firewood from Mount Halimun Reserve, West Java.

BELOW RIGHT Moluccan woman spreading out mace (left) and nutmeg (right) to dry in the sun on Tidore Island. These spices come from the same thick-skinned fruit of the tree *Myristica fragrans*.

BELOW A common sight on Bali is the taking of impressive and beautiful votive offerings to temples.

LEFT Dani man and his sons taking home firewood in the Baliem Valley in the highlands of Irian Jaya.

OPPOSITE PAGE Forests throughout the archipelago are suffering at the hand of man, as here near Manokwari, Irian Jaya, often to the detriment of the local people and the country as a whole.

together as a single nation over nearly fifty years, some of them extremely turbulent, and to the way the Indonesian language, devised as a lingua franca only forty years ago, is spoken by the majority of people throughout the archipelago. The diversity seems to be never-ending; one only has to uncover the existence of a cultural group before it appears that they themselves recognize further divisions and strata. It is inadequate to say that the Indonesian people are made up of Acehnese, Bataks, Minangkabau, Malays, Kubu, Mentawaians, Sundanese, Javanese, Balinese, Dayaks, Buginese, Torajans, Manadonese, Ambonese and Danis, for this misses many proud and historic groups, and ignores the many subdivisions of the main groups.

This is not to say that the groups are, or ever were, totally defined by geography. Buginese from South Sulawesi have traditionally gone to remote shores to literally carve a living from coastal forests around the archipelago. It is a rare town that does not have a Minang trader or Padang-style restaurant with strong ties to West Sumatra, or a Batak, Manadonese or Ambonese in government service. Indeed, it is said that there are more Bataks and Minangs in Jakarta than there are in their

homelands of Sumatra; and the buses leaving Jakarta for Sumatra before the big holiday at the end of the Islamic fasting month are packed with travellers returning home. Possibly the only exceptions to this mixing are the Irianese, who are relatively small in number, have not traditionally travelled far (not for nothing does the island of New Guinea cover only 0.15 per cent of the Earth's surface but have 15 per cent of its languages), and whose province has had an educational infrastructure only in quite recent years.

But the diversity is not just seen in language. As one moves from one end of the country to the other, very little actually stays the same. House design, religion, agriculture, dress and cloth, mode of transport, hunting weapons, script and history change like a slowly turning kaleidoscope.

This is not the place to list and describe the different cultural groups, their prides and joys, their roles in the Indonesian economy, or their relationships with natural resources in the forests and the sea. A traveller exploring the wilder side of the archipelago will encounter many people and, like many before, will be won over by their smiles and friendliness.

Threats to Wild Indonesia

All natural ecosystems contributing to the splendour of wild Indonesia are under great threat, and this applies both to those in reserves and national parks and those outside such 'protected areas'. No single group of people should be singled out for blame in this situation, and in some ways we are almost all culpable since we partake directly or indirectly in the trade of goods which those ecosystems supply, be it timber, plywood, rattan furniture, coral jewellery, aquarium fishes, or food fishes. The threats to forests and coral reefs are considered in more detail below.

Forests

In the forests the logging companies provide easy access to previously remote areas along the roads they build at great expense. Local people and migrants use these roads to hunt the wildlife and there is often no effective prevention by the government because the newly opened regions are, by definition, remote, the areas needing protection are huge, and the available personnel woefully scarce. The logging companies are officially responsible for the integrity of the forest concession whose licence they hold for twenty years, but in reality they have no desire to raise the hackles of the opportunist new-comers and surrounding villages for fear of reprisals and losing cheap labour. As the hunters and others penetrate further into the forest so there is a desire to build at least temporary shelters at the forest edge which, over time, become small settlements. Plots of land need to be cultivated to supply the needs of the families, and a ready market is found for the timber which was too small or of undesirable species for the commercial loggers. In time, a government servant might be stationed there but, by then, the area is no longer remote and the essential problem has moved way up the road. And so the process continues.

But why are the logging companies in the forest in the first place, if not to supply consumer demand? Unfortunately, one of the reasons the demand is so great is that the price of tropical wood in a western market-place needs to be competitive with plantation-produced timber grown nearer to home. But the real ecological value of the tropical forest timber is much greater than that of, say, a plantation softwood. The low value of the timber on the world market has not encouraged a conservation ethic among loggers, and the government offices receiving royalties and fees have been quite happy with the sums coming in from the abundant resource, unaware that they could be even greater. What is often practised is short-term resource 'mining' rather than long-term resource harvesting.

Campaigning literature states baldly that logging leads to erosion, siltation, desertification, loss of species and so on. This is all true but it is not inevitable. All the serious, long-term research indicates that selective logging, conducted according to the rules, only temporarily disturbs the wildlife. But companies

Pockets of forest are cleared and planted with crops, as here in the Padang Highlands near Sibolga in West Sumatra. It is unlikely that this forest will ever regrow.

Once roads are built, so migrant settlers arrive, build a house, clear forest for subsistence crops and then cash crops, while the forest edge recedes further and further into the distance.

will turn blind eyes and take short cuts when and where the rules are inconvenient. Even when (or if) companies do play the game by the book, their efforts can be brought to nothing by the actions of other people clearing and burning the forest; logged forest is much easier to burn and convert into agricultural land than intact forest.

Despite the protestations of some of the more high-profile campaigners, almost all the forests in Indonesia are capable of a log harvest that could be sustained into perpetuity, and this has been demonstrated in a few areas. Of course, there are a number of caveats to such a statement. First, there must be a recognition of the need for total protection of a forest for many decades after logging, so that natural regeneration can proceed. Second, the forest must be logged at a limited intensity and in a way that minimizes damage to the soil and remaining trees. Third, by whatever means, the loggers must have a vested interest in the proper care of the forest. Lastly, the price of forest timber must reflect its true value.

These issues of licence agreements, fee assessments, tax incentives, politics, trade imbalances and export tariffs are extremely complex and the Indonesian government is taking steps to improve the situation in order to back its desire to be seen as a responsible timber producer committed to sustainable development. No one would insist that even the planned measures are perfect, but then nor are world trade and tariffs or debt arrangements, both of which are dominated by western countries.

Coral Reefs

The coral reefs of Indonesia are also facing destruction by greedy exploiters of their riches, who use explosive charges, poisons and diving gear. Coral grows very slowly and the destruction of coral can set a reef back decades in its development. Explosions and poisons are completely unselective in

their effects and many of the creatures killed are not wanted and will not be sold. Even the most remote coral reef is not safe from depredations, since no reef is more than a few days' sail from the last. The problems of reefs are not dissimilar to those of the forests. The manpower, boats and budgets required to patrol the large marine reserves already in existence and planned in Indonesia are way beyond what is currently available. Yet the demand for the products of coral reefs, such as shells, sea cucumbers, aquarium invertebrates and fishes, and for the coral itself for limestone and road building, seems to be increasing.

An example of what has been lost through the use of coral for construction is Ambon Bay in the Moluccas, about which Alfred Russel Wallace wrote nearly one and a half centuries ago 'There is perhaps no spot in the world richer in marine productions, corals, shells and fishes, than the harbour of Amboyna'. If only that could be said today of the wrecked and rubbish-strewn bay.

Conservation of Wild Indonesia

One of the mainstays of any country's conservation efforts is a well-managed system of protected areas containing large, representative areas of all the natural ecosystems. To this must be wedded a constant programme of public awareness, and full attention must be given to the problems and needs of the people surrounding the protected areas. Allowance has to be made for traditional needs of the people, particularly in tribal areas, so that human suffering does not result from the protection policies. At the same time steps must be taken against perpetrators of misdemeanours regarding protected areas and protected species, which require the enactment and enforcement of laws, and the understanding and support of the judiciary. Much of this infrastructure is already in place in Indonesia under the authority of the Directorate General of Forest Protection and Nature Conservation in the Ministry of

ABOVE These young Komodo Dragons (*Varanus komodoensis*) are three to four weeks old. They were hatched at Jakarta Zoo in 1990, a new achievement for the zoo, although Surabaya Zoo in East Java has a tradition of breeding these enormous lizards. Such animals can be traded or presented to foreign zoos without depleting the wild stocks.

BELOW Confiscated Sumatran Orangutans (*Pongo pygmaeus*) are taken to Bohorok Rehabilitation Station on the edge of Mount Leuser National Park where they are rehabilitated to life in the forest. Every few months the artificial feeding is stopped for a while to force these resourceful apes to find food in the forest.

Forestry, but insufficient manpower and budgets frustrate their best efforts.

The problems of Orangutans and elephants are indicative of the complexities of some of the issues facing the government and conservation advisors. In some cases 'ecotourism' is vaunted as a solution to funding a conservation area, but this approach has both important pros and cons. These topics are discussed more fully below.

Orangutan Rehabilitation

Young Orangutans (*Pongo pygmaeus*) are quite delightful; their large dark eyes, their quizzical expressions and rubbery mouths capable of a wide range of human expressions endear them to us, and have been the cause of the death of many adult females, from whose cooling bodies the infants are torn. Over the last twenty years, there has been an ever-increasing awareness of the damage this does to the wild populations and to wildlife conservation in general. The government conservation authorities began to confiscate pet Orangutans in the early 1970s and a need was soon seen for the establishment of rehabilitation stations to enable these animals to be returned to the wild. As the parallel education programmes had their effect, so other animals were handed over, some from high-ranking officials for whom the Orangutan in a cage was a considerable status symbol. The Orangutans were taken to the rehabilitation stations where they were initially put into quarantine because they can contract many human diseases. After this period, the

infants continued to be cared for by the station personnel and the juveniles were released into the forest and left – but only for a while. Twice a day they would be fed bananas and milk, and at other times they were free to explore the forest. Slowly they would adapt to forest life, and they were further encouraged to be independent by the policy of stopping artificial feeding for a week to force certain stay-at-home individuals to go out and find food. These stations have been successful in returning many animals to the forest, which have then bred, and much has been learnt during the process.

Despite this success, it is now generally agreed that releasing relatively tame Orangutans, which may be carrying communicable diseases, into wild populations which may already be at an optimum density, is unwise. Better would be the release of these animals into safe areas of forest without a wild population. But the number of such areas within the natural range of Orangutans is increasingly limited as forests are lost for one reason or another. However, rehabilitation stations remain excellent places to see semi-wild Orangutans and learn about them.

Elephants as Symbols

The Sumatran Elephant (*Elephas maximus*) holds the enigmatic position of being a protected pest. The clearing of large areas of lowland forest in recent years, and the disruption of traditional migration routes, have led to the confinement of elephants within smaller and smaller areas of forest. Elephants have been forced to live at high densities in Way Kambas National Park and Air Sugihan Reserve in southern Sumatra as a result of dramatic elephant drives in surrounding areas organized by the government. These have used everything from army helicopters, fire crackers and amplified music to yelling villagers with flaming torches. The elephants have been able to survive in high densities in these sanctuaries because of large areas of succulent secondary vegetation. There has, however, been a dramatic increase in reports of elephant–human conflicts over the last decade as elephants enter farmlands to feed on young coconut and oil palms, rice, corn, cassava and bananas. In some cases the very success of plantation projects has been put in jeopardy because of the elephants' depredations. Such conflicts must be minimized to protect human life and livelihood while conserving as many wild populations as possible in protected natural forest.

There are two ways of regarding the elephant problem. One is to look for a specific engineering or management solution. Ditches and other barriers can be constructed to confine the elephants, but these are expensive and require considerable manpower to maintain. The other approach is to recognize that the dramatic plight of this, the largest of Asian mammals, is symbolic of the plight of the rest of Indonesia's wildlife. Such problems can be dealt with only by making serious attempts to improve reserves by applying existing policies and regulations with proper budgetary commitments. Until this happens the crop raiding will continue to be dealt with on a case-by-case basis, analogous to treating the symptoms of a disease by expensive surgery, while not tackling its cause.

Meanwhile, the elephants must be kept away from humans by providing essential mineral or salt sources within their forest ranges, by erecting well-maintained solar-powered electric fences between the forests and fields or plantations, or by providing wide buffer zones around the forests with no suitable elephant cover or food. Unlike their African cousins, Asian elephants are unable to stay in the sun during the day and have to seek cool cover under trees. However, it may be essential in the near future to cull the present population judiciously within the precepts of a scientifically based management plan.

Ecotourism

There are some who point to 'ecotourism' as a source of salvation for the richest wildlife areas, based on the model of East African national parks. Unfortunately, the model is inappropriate for a number of reasons, not least the total discrepancy in the visibility and accessibility of the animal life: you cannot drive up to a Sumatran rhino and take its photograph. The main exception to this is probably Komodo where travellers (sometimes hundreds a day) come to see and photograph the huge, wild Komodo Dragon (*Varanus komodoensis*). Interestingly, the most-visited national parks and nature reserves reflect much more the importance of *geo*-tourism, since they tend to be centred on the attractions of dramatic volcanoes, waterfalls and rugged coastlines rather than biological interests. Foreign tourists are more inclined to enter forests than domestic tourists, but the volume of the latter is slowly increasing. Part of the reluctance of domestic tourists to explore the forest stems from a traditional battle against the forest. Pioneers in any area had to beat back the forest to build their houses and establish their fields, and the forest in its turn would grow back if allowed. In addition, most of the ancient, underlying animist religions of Indonesia would have one believe that the forest is an eerie, evil and frightening dwelling-place of spirits.

BELOW Hundreds of domestic and foreign tourists visit this impressive waterfall each weekend in the limestone hills of Bantimurung north of Ujung Pandang in South Sulawesi.

ABOVE 'Help!! – where must we go now to find a home?!!' – a World Wide Fund for Nature poster to promote awareness of conservation problems.

ABOVE RIGHT Whitewater rafting is becoming an increasingly popular tourist activity on Indonesia's wilder rivers, as here on the Alas River winding through Mount Leuser National Park.

Conclusion

Although western schoolchildren are now taught that the tropical forests of the world contain a phenomenal and valuable cornucopia of products such as drugs, industrial chemicals, and untried fruits and vegetables, the reality is not so simple. An economist or a government is forced to argue that while these useful things may be there, and may have already been found in other forests, they just as well might *not* be found in the area of forest earmarked for development into, say, an oil palm plantation which has well-known and foreseeable costs and expected monetary benefits. Simply to locate a possible product is likely to require expensive laboratory space and technicians. Local people may be able to provide short-cuts, but many of the local medicines are as much placebos as they are medically effective. The rigorous testing required before a drug can be marketed is also an enormous and risky expense.

Even the well-tried argument that tropical forests are needed to protect watersheds and so protect water supplies is now becoming somewhat threadbare, as scientists evaluate long-term data and discover that in certain situations a good grass cover is just as effective as forest for soil protection and recharging underground water sources.

It would appear that the only absolute arguments for conservation are in the realms of ethics, morals and religion. The economic arguments for saving the rain forest are best left alone, because an even better economic argument may be forthcoming for an industrial conurbation. There are, however, no ethical, moral or religious grounds for causing extinctions, or for depriving future generations of the spiritual uplift afforded by a walk in a forest or the view of a coral reef. As Dr C.G.G.J. van Steenis, one of the fathers of natural history in Indonesia, replied when asked why one should bother to conserve the forests, 'Come with an observant eye and the question will answer itself'.

Sometimes the plight of natural ecosystems may seem rather depressing, but it is not too late, there is still a great deal of forest and some exceptionally beautiful reefs. The conservation message is being trumpeted with increasing volume in Indonesian newspapers, magazines and on television. In addition, there is a growing movement of concerned young and not-so-young people within the country.

Sumatra and Kalimantan

Sumatra is the westernmost major island of Indonesia, and Kalimantan is that part of the island of Borneo which is administered by Indonesia. Both are renowned for their wild forests; here are the giant trees, the lianas as thick as one's leg, the exotic flowers, the large and beautiful but dangerous animals, the mighty rivers, the deep swamps, the lofty mountains – the very quintessence of the word 'jungle'. Sumatra and Kalimantan have much of their natural history in common and many of their species are found on both islands. This is not surprising given their proximity and the similarities between their geological histories which saw them separated by the watery barrier of the Karimata Strait only about 7,000 years ago.

The flora of Borneo is the most diverse in the region. It has some 11,000 species of flowering plants, about a third of which are endemic, and there are about sixty endemic genera from various families. Over 700 species of trees have been recorded in just 6.6 hectares (16 acres) of lowland forest; for comparison, there are only fifty native tree species in Europe north of the Alps and west of the USSR, and 171 species in eastern North America. Sumatra is less species-rich than Borneo, and only 12 per cent of its plant species and seventeen of its plant genera are endemic. The fauna of Sumatra is, however, slightly more diverse than that of Borneo, perhaps because it is slightly closer to the Asian mainland. Many of Borneo's animals are exclusive, with thirty-nine land mammals and thirty birds being endemic, compared with only nine and thirteen respectively on Sumatra.

Coasts

It is clear from a brief glance at a map that Sumatra and Kalimantan have a variety of coastlines. Kalimantan and eastern Sumatra are crossed by many large rivers which flow into shallow and relatively sheltered seas through broad estuaries and spreading deltas, carrying large quantities of silt. Many towns have grown up on these rivers which are transport lifelines into the interior. In contrast, on the west of Sumatra there are few towns of any size and the rivers are small, flowing into the deep Indian Ocean where wave action is strong. As a result the eastern shores have extensive mudflats and wide tidal mangrove forests, whereas the west has rocky headlands and sandy bays where the tides are less pronounced.

The mudflats of these sheltered eastern shores may appear quite lifeless on a cursory visit, but there is a considerable amount of activity both on and below the surface fuelled by the high organic content of the mud. This comes from the debris brought down by the rivers or dropped from the mangrove forests. The animals living here are decidedly marine with crabs, snails and polychaete (bristle) worms being the most common. Right at the edge of the water, mudskippers can be seen, more often than not with just their bulging eyes above the water surface. These unlikely looking fish can walk, run and even jump on the surface of the mud using accomplished flicks of their tails. They can even climb up sloping trees using their pelvic and pectoral fins. Some species feed exclusively by grazing on the thin layer of green alga on the mud surface and others use their agility to supplement their diet with small crabs, while some have an even broader diet which includes insects, snails and other mudskippers of lesser agility. They live in burrows, some of which open into mud towers and others into depressions surrounded by pellets of mud thrown up by the fish while maintaining the burrow.

The exposed rocky headlands of western Sumatra have Screwpines (*Pandanus*) and other beach trees such as the needle-leaved She-oak (*Casuarina equisetifolia*), the yellow-sapped *Calophyllum* and the feathery-flowered *Barringtonia* clinging to them. Lower down, the plants and animals are being pounded remorselessly by waves breaking over them one moment, while the next they risk being desiccated as the tide falls and the sun bakes the dark-coloured rock. Some of the time they are entirely covered by salty sea-water while at other times they are being flushed by pure rain-water during a tropical storm at low tide. Not surprisingly, the wildlife needs special adaptations to survive in this hostile environment. A wide variety of invertebrate groups is represented and all of them have means of maintaining a secure hold on the rocks; those that cannot move easily, such as barnacles, have developed means of shutting themselves off from the world when conditions are unfavourable. The reaction of other animals is to hide in crevices or to move up and down with the tides.

Lowland Forests

Both Sumatra and Kalimantan have a wide range of physical conditions influenced by altitude, soil type and slope, with the result that there is an equivalent range of forest types, each of which has a distinct ecology.

Swamps

Few areas of freshwater swamp remain in Sumatra and Kalimantan, the best known being Way Kambas National Park in southern Sumatra, but even their combined area is now insignificant because they have been found to be extremely suitable for conversion to agricultural land. Down the eastern side of central and southern Sumatra and southern Kalimantan are large areas of swamp forests. On the seaward edge these are the mangrove forests, but on the inland edge they are peat swamps.

The surface of most of the extensive peat swamps of Sumatra and Kalimantan is above that of the surrounding land so that no nutrients enter the swamps from rain draining through or over the surrounding land. The swamps are, therefore, a closed system and the plants have to use nutrients from within themselves, the peat, or the rain that falls directly on them, with the result that the peat is very acid and poor in nutrients.

The soil of a peat swamp comprises a crust of fibrous, dead material tied together with the roots of living plants. Below this is a semi-liquid layer with wood or other vegetable matter suspended in it, and at the bottom is a mat of heavy debris overlying a mineral soil. This soil used to support mangrove forest up to 8,000 years ago, but the mangrove fringe has moved with the coastline as the sea level has fallen. The mangrove soil left behind had such a high sulphur and salt content that decomposer organisms were unable to perform effectively, with

ABOVE The mangrove-fringed shores of the estuary of the Barito River in South Kalimantan may be muddy but they support a wide diversity of wildlife from fishes and prawns to birds and reptiles.

BELOW Mudskippers breathe through gills like normal fish when they are under water, but they can stay out of water for short periods by holding water in their gill-chambers, and are also able to 'breathe' through their skin and fins.

BELOW Before surveys began a few years ago, the total world population of the Milky Stork (*Mycteria cinerea*) was believed to be about 1,000 individuals distributed between southern Vietnam and Bali. It is now estimated that 6,000 can probably be found on Sumatra's eastern shores alone.

LEFT The water flowing out of a peat swamp, such as here by the Sekunir River, Central Kalimantan, is generally crystal clear, seeming black when seen from above and dark brown when looked at in a glass. Many very beautiful species of fishes, some of them popular in aquaria, are found in these waters.

OPPOSITE PAGE, LEFT Heath forest or *kerangas* on Bangka Island, South Sumatra. This type of forest is characterized by relatively small, short trees growing on sandy soil, and although there is little left on Sumatra, there are still large areas remaining in Central and East Kalimantan.

the result that plant material falling to the ground remained undecomposed, forming an ever-increasing layer above the soil.

The forest of a peat swamp generally occurs in up to four rings of different types, the trees of which become progressively shorter and more stunted towards the middle, where the peat is deepest. The trees in this so-called *padang* forest in the centre also tend to have thick leaves, and many of them are from the myrtle family such as *Tristania obovata*. In the outer peat swamp the forest is similar to lowland forest but has fewer species, and there are some good timber trees such as various species of the dipterocarp *Shorea* and, the most valuable of all, the Ramin (*Gonystylus bancanus*). Another tree, the poisonous *Gluta renghas*, is quite common and has a fine red timber, but its latex causes blisters on contact with skin so it is not often felled. Over twenty species of palm grow in a rich peat swamp forest, and the following three are quite common. The most intimidating of them is the massively thorny Wild Salak (*Salacca conferta*), a relative of the Edible Salak (*Salacca zallaca*) whose scaly, astringent fruit are commonly seen in markets and shops. The most conspicuous and tallest palm is the Serdang (*Livistona hasseltii*) which is often left standing when a peat swamp forest is cleared, because its extremely hard bark makes it very difficult to chop down. Quite the most beautiful palm is the Sealing Wax Palm (*Cyrtostachys lakka*) the leaf bases and central ribs of which are bright red. Young specimens command a very high price and can be seen with sack-bound roots perched on the top of long-distance buses or even arriving in Jakarta airport among the suitcases.

Animals are scarce in peat swamp forests, with some monkeys living at only a third of the density they do in inland forests. No large animal is confined to peat swamps, although the small, fish-eating crocodile known as the False Gharial (*Tomistomus schlegeli*) favours this habitat, and some of the small and rather attractive, air-breathing gourami-type fishes are quite common in the black waters.

Heath Forests

Kalimantan has the most extensive areas of heath forest in South-east Asia, while only a few small patches remain in Sumatra where it was never very extensive. Heath forest is also known as *kerangas*, a graphic name meaning 'forested land which if cleared will not grow rice' in the Iban language of interior Kalimantan. Heath forest usually grows on soils from silica-rich rocks which are inherently poor in nutrients, and sandy textured, through which water drains easily. The soils are often referred to as 'white-sand soils' which are covered with a thin layer of dead leaves and other organic material. Below the white sand, perhaps a metre below the surface, a hard, iron-rich layer called a podzol pan develops which can stop the free flow of water, causing flooding after heavy rain. The water flowing out of a heath forest is usually a blackwater, as in peat swamp forest, full of dark humic acids and with very few nutrients.

The heath forest itself has a low, uniform, single-storey canopy formed by large saplings and 'pole' trees which are so dense that walking through the forest can be quite difficult. There are only half the tree species present that would be found in a normal inland dipterocarp forest (about 100 instead of 200). Many of the features of heath forests have parallels in the forests of high mountains and peat swamps, both of which also have poor, acid soils; in fact, many of the heath forest tree species are also found in peat swamp forest. Where heath forest has been felled or burned, a form of slow-growing, stunted *padang* vegetation forms with the fine-leaved *Baeckia frutescens*, the Paperbark (*Melaleuca cajuputi*) and the fire-resistant, thick-barked *Fagraea fragrans* as characteristic members. It appears that once heath forest is felled, it cannot regrow because the organic layer of the soil is quickly oxidized, and what little clay there is within the soil is likely to be washed down rapidly through the sandy soil. This causes the now notorious fragility of these soils; unfortunately this was not understood before Javanese trans-migrants were moved to new settlements on apparently fertile land that had supported heath forest.

ABOVE Lowland dipterocarp forest between Balikpapan and Samarinda in East Kalimantan.

Both *padang* and heath forest characteristically have numbers of plants which have special relationships with ants and other insects. Some of these are epiphytes such as the spiny 'ant-plants' *Hydnophytum* and *Myrmecodia* which provide protection for the ants within their swollen stems. In addition, insectivorous plants such as Pitcher Plants (*Nepenthes*) and Sundews (*Drosera*) are commonly found. All these life forms are examples of adaptations to the low nutrient environment. This harsh environment does not seem able to support much wildlife, perhaps because the tree species that can live in the relatively extreme conditions do not produce adequate or sufficiently attractive fruit, seeds, flowers or leaves to support many species of animals.

Dipterocarp Forests

Attempting to tread silently through dipterocarp forest while stalking animals can be exasperating as enormous crispy leaves, some of which may be as large as a dinner plate, crackle underfoot. These leaves are probably from dipterocarp trees, members of the family Dipterocarpaceae. This is one of the largest families of forest trees anywhere in the world; they are typically large trees making up the main framework of the forest. This family is almost entirely confined to South-east Asia, with a single species in South America, about forty species in West Africa, but about 470 species from the Seychelles to New Guinea, most of which are found in lowland forests. Borneo has by far the most species with 267, 155 of them endemic, Sumatra has 106 species, only eleven of which are endemic. As one goes eastwards, so the number of species per island decreases, with only ten on Java, seven on Sulawesi, six in the Moluccas, and fifteen on the island of New Guinea (relatively few given its large area).

Many of the dipterocarps are huge, some reaching 75 metres (250 feet) in height and with a diameter above the buttresses, the woody 'supports' at the base of the tree, of over 2 metres (6 feet). Their size, together with the characteristics of the wood of many species, have made the dipterocarp trees among the most favoured timber trees in the region. Regeneration of the saplings and seedlings is such that, if managed properly, repeated harvests can be taken of the timber, and there are a number of systems devised to achieve this. Unfortunately, supervision is often lacking, with the result that the pressure of harvesting cannot be sustained and irreversible degradation of the forest occurs.

The valuable timber is used primarily for light construction, furniture, plywoods and veneers. Other useful products include fat from the species of *Shorea* known as Tengkawang, resins from species of *Dipterocarpus*, and camphor for perfumes, incense and embalming from species of *Dryobalanops*. This camphor is a very ancient article of trade from Sumatra, and was mentioned by Marco Polo in 1299. At the village level, dipterocarp bark and leaves are used for thatch and walls, and the wood is converted into charcoal, although these practices are rarely seen nowadays.

Seasons

On successive visits to an area of lowland forest it will be noticed that the numbers or proportion of trees flowering, fruiting or with new flushes of leaves changes and there does not seem to be much of a predictable pattern. Some species flower and fruit at the same time, others do not; some follow an annual cycle, others follow longer or shorter cycles; a very few species, such as certain figs, fruit at regular intervals without regard to weather or season, whereas others seem to respond to some environmental cue; some produce new leaves all year round, others produce them twice a year, while still others produce them in a rush after an 'autumnal' leaf fall. Some trees produce flowers which always develop into fruit, whereas other species

FAR LEFT The wild Sugar Palm (*Arenga pinnata*) (photographed here in Kerinci-Seblat National Park, Jambi) is one of the few trees which fruits all year round. The flesh around the seed contains a high concentration of oxalic acid crystals which decreases as the fruit ripens, thereby discouraging animals from eating it before the seed is fully developed. In this way the tree ensures that its seeds are mature before they are dispersed.

LEFT Dipterocarp seeds are high in food value, being rich in oil and with few toxic chemical defence compounds. The seeds of some *Shorea* species contain 70 per cent fat, similar to cocoa butter, and are collected, processed and exported for use in chocolates and cosmetics. Forest-based people have traditionally boiled the seeds and used them as vegetables.

seem to have difficulty producing fruit even though they may flower regularly. The cues causing flower, fruit or leaf production in different species can only be elucidated by observations made over a long period.

Most commonly, leaves are produced twice each year with the major peak occurring just after the driest period; flowering seems often to be initiated by water stress, and thus peaks just after the driest time of year; while fruiting peaks about five months later, just before the wettest time of year. At any one time there are more fruiting trees than flowering trees because the fruit persists on the trees for longer than the flowers.

The changes in the abundance of flowers, fruit and young digestible leaves clearly have an influence on many facets of forest life. Many trees are pollinated by insects which are most common during the flowering periods. Leaf-eating caterpillars become abundant after new leaves are produced. Other insects, such as those dependent on rotting wood or on pools of water for breeding, are most abundant during the wetter months. These changes in insect abundance also influence the behaviour of insectivorous animals which have restricted ranges, such as many species of birds. Insects are also a very important supplementary food for birds which mostly eat fruit, because they provide the high-value food necessary for energy-draining activities such as breeding and moulting. Consequently breeding and moulting in these bird species peak with increases in insect availability. Fruit abundance clearly influences frugivorous animals, with gibbons, for example, responding to fruit abundance by travelling widely around their well-defined home ranges, and forest rats responding by breeding at such times.

Of all the flowering and fruiting cycles, the most bizarre is that of the dipterocarp family, which has simultaneous mass fruiting over wide areas at intervals of about two to seven years, triggered perhaps by occasional severe droughts. Such a pattern is unknown outside South-east Asia. The primary result of this mass fruiting is to allow the seeds to escape predation by beetles and other animals. This is not to say that fruit will not be eaten, but rather that the predators will be so satiated by the enormous and sudden production of seeds that the majority of the crop will escape unscathed. The long intervals between fruiting mean that none of the potential seed predators will be able to develop specializations to predating on the dipterocarp seeds, and their populations will remain at a low level until just after fruiting begins. The wonderful feature of this behaviour is that it works

only because virtually all the family members behave in the same way – a single species fruiting gregariously would probably reap no benefit. Interestingly, while fruiting is synchronized, flowering is not, with some fruit taking longer to develop than others. This probably has considerable advantages because many of the pollinators, small flying insects called thrips, pollinate a number of species, and if all the species flowered at once there would not be enough thrips to go around.

Mistakes do occur, of course, and it is not unknown to find just a few dipterocarp trees fruiting in isolation in an area, perhaps where a local drought has occurred. Trees will sometimes correct such a mistake by absorbing or aborting flowers and flower buds, or producing sterile fruit, thereby minimizing their losses. Any fertile seeds which are produced will almost certainly be found and destroyed by predators.

Death and Decay

Sit still in any lowland forest, and it is not long before one senses the dynamic nature of the place. Branches large and small drop from the canopy with surprising regularity, old trees crash to the ground with little encouragement or warning, and the constant drizzle of leaves is vastly accentuated in the wind before a storm. If one tries to lift a dead leaf delicately off the forest floor, several leaves will come away together, being bound together by fungal strands which are absorbing nutrients from the leaves. Just below the rotting leaves, one can find fine roots from trees and other plants absorbing nutrients released from the organic litter. In fact, many plants have special fungi within or around their roots. Shortly after dusk, when no celestial light reaches the forest floor, ghostly, whitish-green shapes can be seen on the fallen leaves, which seem to vanish mysteriously if one looks too hard. These are luminous patches and strands of fungus weaving through and digesting the dead leaf litter. The mushroom-like fruiting bodies of certain species also glow in the dark, some with such intensity that it is possible to read a book by their light. To appreciate another aspect of the forest disposal system for organic matter, drop a few grains of rice or fragments of meat onto the forest floor. In less than a minute an ant will probably have found the food and either spirited it away, or have gone to tell members of its nest about the unexpected bonus and the need for recruits to be assigned to the removal gang.

Within the forest soil are vast armies of termite workers which play a very significant role in the important processes of decomposition and nutrient cycling. Although termites may look like ants, they are actually much more closely related to cockroaches. They live in large communities with a million or more members, but these members are not strictly different individuals because all except the parents, the 'royal pair' of king and queen, are identical siblings, some of which develop as workers and some as soldiers. A patch of good lowland forest may contain over fifty species of termites, although certain species dominate either in terms of the number of nests or the size of colonies.

On the night after the first rain ending a dry period, millions of winged termites swarm from their nests in the ground, providing a food bonanza for hosts of birds, lizards, frogs and, in some communities, people. They will lose their wings when they land and a male and female seek a hole in the ground or a crack in a tree where they establish their 'royal cell'. They mate and the female's body starts to swell as she begins the last phase of her life, that of being a sausage-shaped egg-producing machine. Some termites develop into blind, sterile workers, others into ferociously jawed soldiers which protect the workers when they leave the nest, or guard the entrance to the nest.

Most of the animals involved in the forest decomposition processes depend directly or indirectly on bacteria or fungi to perform an initial processing job on wood and leaves, since no animal is able to digest the chief constituent, lignin, and few, not even termites, can cope with cellulose. One group of termites gets around this problem by cultivating gardens of fungi in their nests which process the lignin and cellulose for

ABOVE Fungi (possibly a species of *Mycena*) in Kutai National Park. Fungi play a vital role in recycling nutrients within the forest by breaking down dead plant and animal matter, making it available to plants and small animals.

LEFT Landslips, such as this in Kerinci-Seblat National Park, are entirely natural phenomena and certain plants specialize in colonizing the bare earth, preparing it for larger trees in their turn.

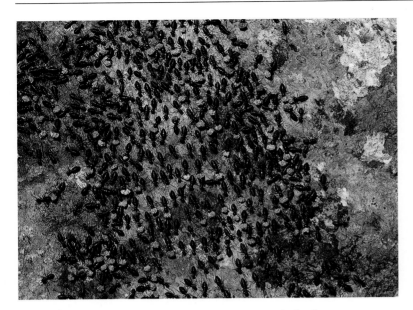

Columns of termites can be seen marching through the forest, up trunks, along the forest floor and into holes in the ground, carrying little pellets of wood chewed from dead trees. Some species employ special fungi to help break down the wood so that the maximum food value can be obtained.

The cabbage-sized bud of *Rafflesia gadutensis* at Batang Palupuh Reserve in the Padang Highlands of West Sumatra. The plant is a parasite and has no leaves, no stem and no proper roots. It will probably be a few weeks before this bud opens fully.

them, producing a more digestible, richer food. Not surprisingly, these termites are enormously successful and it has been estimated that about one-third of the leaf litter in a forest is removed by them.

The major termite predator is the Pangolin (*Manis javanica*). This armour-plated insectivore has a long, sticky tongue, and strong, curved claws to rip open termite nests. A dedicated researcher once counted about 200,000 termites in the digestive tract of a Pangolin. If one assumes that this is a Pangolin's average daily intake, then a single animal may consume something in the region of 73 million termites each year. Many other animals, such as gibbons, also take termites in small quantities when these insects make forays in serried ranks across the forest floor and into the trees.

Giant, Stinking Plants

Most famous of all the plants in the lowland forests of Sumatra and Kalimantan is one popularly known by its scientific name, *Rafflesia*, commemorating the discoverer of the first and largest of the several species, the flamboyant and controversial Sir Stamford Raffles. He found the plant while on his first excursion into the hills behind Bengkulu in south-west Sumatra at the start of his five years as Lieutenant-Governor there between 1818 and 1824, during which time he also founded Singapore.

In a letter to his friend and patron, the Duchess of Somerset, on 11 July 1818, Sir Stamford wrote:

The most important discovery throughout our journey was made at this place; this was a gigantic flower, of which I can hardly attempt to give anything like a just description. It is perhaps the largest and most magnificent flower in the world, and is so distinct from every other flower, that I know not what to compare it – its dimensions will astonish you – it measured across from the extremity of the petals rather more than a yard, the nectarium was nine inches wide, and as deep; estimated to contain a gallon and a half of water, and the weight of the whole flower fifteen pounds. This gigantic flower is a parasite on the lower stems and roots of the *Cissus Angustifolia* of Box [the vine is now known as *Tetrastigma*] . . . The whole substance of the flower is not less than half an inch thick, and of a firm fleshy

consistence. It soon after expansion begins to give out a smell of decaying animal matter. The fruit never bursts, but the whole plant gradually rots away, and the seeds mix with the putrid mass.

Precisely how the seeds are dispersed is unclear, but it may be that mammals such as rats, tree shrews, mouse deer and pigs, or birds such as junglefowl or pitta, scratch around in the dead flowers and get seeds stuck on or in their feet. If the seeds are then rubbed off against the bark of the *Tetrastigma* vine, they will germinate and grow. This may seem a long chance, but the cards are further stacked against *Rafflesia* because each flower is either male or female. Thus flowers of both sexes must be open at the same time for flies and small beetles, attracted by the smell, to effect pollination.

Less well known, but just as remarkable, is the enormous flower structure of *Amorphophallus titanum* discovered in West Sumatra sixty years after *Rafflesia arnoldi* by the great Italian explorer-naturalist Odoardo Beccari. Respected though he was within the scientific establishment, the measurements he gave of the plant were such that people in Europe had to see it for themselves before they would believe it. That opportunity came in 1889 when the tuber, about 2 metres (6 feet) in circumference, which had been growing at London's Kew Gardens for some years, finally produced a bloom 2 metres (6 feet) tall. The tallest bloom recorded for this species is 3.3 metres (11 feet).

Amorphophallus titanum is a giant aroid lily, a relative of lords-and-ladies and taro, and is generally seen as a blotchy stem up to 20 centimetres (8 inches) across, sporting intricately convoluted leaves. These grow and die away over a number of years, becoming progressively larger, until the tuber growing beneath the ground is big enough to produce the flower structure. The largest tuber ever measured weighed 75 kilograms (165 pounds). Even less well known than *A. titanum*, but only a little less remarkable, is its relative *A. decus-silvae*, first described only seventy years ago, which has an inflorescence and stalk which together reach up to an astounding 4.4 metres (14 ½ feet) above the forest floor, but the flowering part itself is only 1.5 metres (5 feet) tall.

If one wishes to award superlatives to these plants, then *Rafflesia arnoldi* has the world's largest single flower, *Amorphophallus titanum* has the largest unbranched inflorescence, *A. decus-silvae* has the largest unbranched flowering structure, but the Talipot Palm (*Corypha umbraculifera*) has the largest inflorescence of all: 6–8 metres (20–25 feet) tall with many branches and some 60 million flowers.

Montane Forest

Mountains are exciting places at any time, but those in the tropics have an added fascination. Just the contrast between the luxuriant, hot and humid lowlands and the stark, depauperate mountains is very striking. Climate and soils change with altitude and the plants and animals respond to this, forming tiers of communities, the highest of which resemble those found in temperate areas.

Frosts occur occasionally on the highest Sumatran peaks, as evidenced by grey, frost-bitten and brittle plant shoots. During the day, montane temperatures can be as high as in the lowlands, but reduced air pressure, low density of water vapour and generally clearer air mean that mountains cool rapidly when the sun is obscured or below the horizon.

The trees at high altitudes tend to be stunted with small leaves and there are few climbers, orchids or ferns. By the time one

Some of the most common species of trees in the montane forests of Sumatra and Kalimantan are relatives of the oaks. This species from Mount Leuser National Park in northern Sumatra is possibly *Lithocarpus korthalsii*.

reaches about 3,000 metres (9,800 feet), the canopy is only a few metres tall, the trees have no buttresses and very small, blunt leaves. Climbers and creepers are absent or very rare but carpets and garlands of mosses and lichens have become common. Near the top of many mountains the vegetation is scarcely over a metre tall and grasses and similar plants predominate. In some cases this is caused by waterlogged soils which are anathema to the high-altitude trees, but in most cases it reflects a history of human interference. For centuries, people have set fires in these areas to increase visibility, thus facilitating the hunting of wild animals such as rhino, mountain goat and deer.

Mammals

Sumatra and Kalimantan have all the mammals that one would expect in an Asian rain forest or jungle – from elephants to large cats, rhinos to squirrels, and monkeys to bears. They have 196 and 221 species of land mammals respectively, about half of which live on both islands. Mainland Sumatra, because of its proximity to the Malay Peninsula, has just nine endemic mammal species whereas Borneo has thirty-eight, ten of which probably do not occur in Kalimantan. Borneo is the richest island in the region for primates, with thirteen species. These range from the diminutive Tarsier and ghostly Slow Loris, to the massive Orangutan and six species of leaf monkeys. The main island of Sumatra can boast only eight primate species, although there are a further four species on the Mentawai Islands to the west. Five of the Kalimantan endemics are tree-shrews, primitive squirrel-like relatives of the primates, and there are four endemic monkeys, nine squirrels and five rats. There are at least eleven species of mammal in Sumatra which are restricted to the mountains and about which virtually nothing is known. These include small shrews, *Crocidura* spp., the Serow or Mountain Goat (*Capricornis sumatraensis*), the shaggy Mountain Giant Rat (*Sundamys infraluteus*) which is over 60 centimetres (24 inches) long from nose to tail, and the attractively striped Sumatran Rabbit (*Nesolagus netscherii*) found only around Mount Kerinci. The other Sumatran endemics are mainly rats and bats, but there is also one monkey, Thomas' Leaf Monkey (*Presbytis thomasi*). Some of the most interesting species are described below.

Elephants

The Asian Elephant (*Elephas maximus*) is found from India, Sri Lanka and Nepal through to Sumatra; its occurrence on Borneo is problematic. It may have been introduced to the north-east by the Sultan of Sulu several hundred years ago, and the lack of fossils, absence of local names, and its very restricted distribution on this large island, would support this theory. However, its distribution happens to coincide with the island's richest soils, abundant mineral licks (very important for the physiology of a large herbivore) and the lowest density of indigenous peoples. Only if fossils are found will we know for sure that it was not introduced.

Asian Elephants frequent lowland forests, but they also venture into mountainous zones. They live in herds or extended families of five to twenty animals led by an old female. Breeding males only enter the herd to mate with receptive females; at other times they either live alone in the forest or in small single-sex groups. Elephants mainly eat the succulent leaves or stems of plants such as bananas, gingers and young bamboo shoots, of which they need some 150 kilograms (330 pounds) each day. Elephants are also partial to fruit and will shake or even knock down a tree to reach it. An elephant herd generally

LEFT Asian Elephants (*Elephas maximus*) are distinguished from the larger African species by their highly curved backs, smaller ears, and the lack of tusks in the female. They have the peculiar distinction of being both a protected species because they are threatened by loss of habitat but also classified as a pest because they raid village crops and plantations.

BELOW The Sumatran Rhinoceros (*Dicerorhinus sumatrensis*) is found in many of the forest patches in Sumatra, but the major population is in Mount Leuser National Park. It makes muddy wallows using its feet and front horn, partly to keep cool during the heat of the day and partly to dislodge parasites from its skin which it scrapes clean on tree trunks. Such 'scraping' trees are a clear sign of the presence of rhinos.

leaves a feeding area looking devastated. However, the scattered leaves and branches, and trampled saplings, are useful as food sources for other large herbivores such as tapir and deer, enabling them to feed on plants which they could not previously reach. At least as important to the elephants as food, and the 70–90 litres (15–20 gallons) of water they need each day, are sources of minerals, which are generally lacking in the main constituents of their diet. Several earth 'licks' are usually visited for different minerals. In some places veritable caves have been dug out by generations of elephants using their tusks or toe nails to get at the earth.

As Indonesia's largest animal, the elephant faces serious problems with its human neighbours, and these have been described in the Introduction to Indonesia.

Rhinoceroses

Two species of rhinoceros, the one-horned Javan (*Rhinoceros sondaicus*) and the two-horned Sumatran (*Dicerorhinus sumatrensis*), have been recorded on both Sumatra and Borneo. The Sumatran Rhinoceros frequents deep forest where it feeds on woody plants. In contrast, the larger Javan Rhinoceros, now extinct on both islands, preferred the rather succulent vegetation that grows in disturbed areas and so it often crossed the paths of people. Unfortunately, it resented human intrusion and made itself a target of man's aggression. The last Javan Rhinoceroses on these islands were shot in South Sumatra

between 1925 and 1928, and the only remains from Borneo are from cave deposits in south-east Sabah dating from 10,000 years ago.

The Sumatran Rhinoceros would rarely come in casual contact with man, preferring forests in the lowlands, hills and even mountains – we have found footprints at 3,300 metres (10,825 feet) in Mount Leuser National Park. It has been recorded from the Chittagong Hills of Bangladesh to Sumatra and Borneo, but never in Java, suggesting it arrived in Indonesia after the rise in sea level isolated Java. It used to be found throughout Borneo, but it has not been confirmed in Kalimantan since the 1930s. Small populations of five to ten animals remain in Malaysian Borneo, Peninsular Malaysia and Vietnam, but it is probably only on Sumatra that there is even a faint glimmer of hope for its long-term future, and that can be kept alive only if the regulations controlling hunting and forest destruction are enforced. The possible extinction of Sumatran rhinos in the wild has important ramifications for the forest ecosystem as a whole: certain trees produce seeds which must pass through a rhinoceros gut before they can germinate. Such trees may include the many wild species of mango. The loss of the rhinoceros would, therefore, eventually cause the extinction of these wild mangos, and in turn the extinction of the species which depend on them. This may have happened already in parts of Sumatra and Kalimantan. In addition to fruit, rhinos eat leaves, particularly the rather fleshy leaves of plants growing in forest clearings and river banks, as well as bark. Sumatran rhinos have been observed to eat so much of the tannin-rich bark of the mangrove tree *Ceriops tagal* that their urine was stained bright orange; tannin is an ingredient in some anti-diarrhoea drugs and it may be that such food is sought by rhinos for medicinal reasons.

Adult male rhinos live in more or less exclusive ranges within which single females, with any dependent calves, live in rather smaller, overlapping ranges. The male attempts to maintain exclusive breeding rights over these females and communicates this to other males by regularly visiting dung or urine sites.

Likewise the sexual receptivity of a female is communicated by the smell of her urine which is sprayed onto leaves at rhinoceros nose height. Even in a pristine area, these large animals probably live at a density of only one every 100 square kilometres (40 square miles) and they are suffering terribly from forest loss and hunting pressure.

Tigers

The Tiger (*Panthera tigris*), sleek, powerful and beautiful, is without doubt the King of the Sumatran forests. It is not found on Borneo and probably never was. The Sumatran Tiger is quite small in relation to the other races, and quite diminutive compared with the enormous Siberian Tiger. Even so, it is a deadly predator and large enough for most people. It is not unusual to read in newspapers of villagers being killed or carried off by tigers, the prime victims being children and squatting women, who look small and manageable. Even when standing up in a forest at night, it is difficult not to shine a torch behind one occasionally in the dread of catching the red reflection from a pair of large, intent eyes. Yet we have been asleep in the forest and woken in the morning to find tiger footprints only a short step from our heads, the only sign of the nocturnal visitor being a chewed pair of rubber sandals. While not strictly nocturnal, tigers do most of their hunting by night, and it is said that they may travel up to 30 kilometres (20 miles) in a single night searching for food. They have catholic diets, taking anything from fish and insects to deer and, most commonly, pigs.

Sumatran Tigers face acute problems. Being predators they live at very low densities – typically one every 50 square kilometres (20 square miles) in relatively undisturbed habitats with adequate prey such as pigs and deer. They can be found in all forested habitats, from lowland swamps to the high altitude plateaux in Mount Leuser National Park. A male's range overlaps the ranges of the females and juveniles and he will try to keep exclusive breeding rights over the two or three females within his range. Thus, to maintain a reasonable population of,

The Tiger (*Panthera tigris*) is the Asian King of the Jungle. Strong and sleek, it prowls through the forest at night looking for likely prey, which may be small animals or deer and pigs. The tigers' status is desperate in many areas of Sumatra as they have suffered from habitat loss and excessive hunting. Man-eaters do occur, but they are a rarity.

say, sixty animals comprising fifteen males and forty-five females, at least 750 square kilometres (300 square miles) would be needed, and there are fewer and fewer such large areas of forest remaining in Sumatra. It is instructive that none of the islands around Sumatra has any history of tigers, except the largest, Bangka, in the south-east. The other islands are simply not large enough to support a viable tiger population. Patches of forest in a sea of agricultural land are as much islands to many animals as land surrounded by water.

Many tigers are trapped and killed each year for their skins, and villagers are more than happy to help the hunters. The monetary reward they receive is probably derisory in relation to the final cost of the skin, but in many areas it is ample reward. Hunting tigers is, of course, strictly illegal, but the hunts occur, by definition, far from the furthest arms of government and mutual complicity more or less guarantees silence should any investigation be made. Not many years ago a cache of dozens of tiger skins was found near Bukittinggi in West Sumatra, revealing that the desire of some business executives to be flanked at their desk by a stuffed tiger with bared teeth has not been entirely shamed out of existence. In addition, there are stories circulating of rich businessmen from within and outside Indonesia who will pay handsomely for the 'privilege' of shooting a tiger.

Sun Bears and Other Carnivores

The Sun Bear (*Helarctos malayanus*) is the smallest of the world's bears and is found in forests from Burma and Thailand through to both Borneo and Sumatra. It may be relatively small, but it is more feared by villagers and hunters than any other forest inhabitant. This bear is rather short-sighted and may suddenly come across a person at very close quarters and be startled. It appears to be completely unpredictable and will charge without obvious warning or cause, rearing up on its hind legs and lashing out with its vicious, long, curved claws. It is these same

The Binturong or Bear Cat (*Arctictis binturong*) is the largest of the civet cats and shares with the Scaly Anteater or Pangolin the possession of a prehensile tail which provides a better grip in the forest canopy. The Binturong is omnivorous, eating some small animals as well as fruit.

claws that leave characteristic gouges up the smooth bark of very tall Sialang (*Koompassia*) trees whose broad boughs often support massive honeycombs which prove so tempting to the bear. Presumably the bear is quite unable to see the combs high above the forest floor, but it can probably smell the honey.

Sun Bears may occasionally be seen during the day when they forage for a wide range of fruit, succulent buds, various insects and other invertebrates. They can also fall foul of man by being partial to the hearts of coconut palms which contain a white 'cabbage' of developing young leaves. Raids by Sun Bears have seriously damaged both smallholdings and larger plantation concerns.

In addition to the bear, the forests of Sumatra and Kalimantan are home to a range of other carnivores including a total of eight cats, six otters, thirteen civet cats and mongooses, one of which, Hose's Mongoose (*Herpestes hosei*), is so far known from only a single specimen. The largest of the civet cats is the prehensile-tailed Binturong or Bear Cat (*Arctictis binturong*), which can reach 14 kilograms (30 pounds) in weight and 2 metres (6 feet) in total length. Almost all of these carnivores favour undisturbed forest and are nocturnal, shy, difficult to observe, and therefore very little known.

Orangutans

In prehistoric times, Orangutans (*Pongo pygmaeus*) could be found from China right through to Java, and these early animals were much larger than those living today. Nowadays they are found only in parts of Sumatra and Borneo. Cave deposits containing Orangutan bones have been found in West Sumatra and it is likely they were once present throughout the island but some factor, possibly hunting, caused their local extinction and now they are not found south of Lake Toba.

Adult male Orangutans can weigh 90 kilograms (200 pounds) and standing up can easily touch the head of a tall man. They have grotesque cheek flanges and their enormous but quiet strength is most impressive. Despite their immense size, Orangutans are completely at home in the tree tops as they thoughtfully consider crossing between the crowns of trees using their long arms and hooked hands and feet, never taking risks but brave enough for acts that a ground-based observer would regard as foolhardy. If the gap between two trees is too great, they will use their weight to sway the tree they are in backwards and forwards until the next tree can be reached. Orangutans only rarely come to the ground, although the largest adult males do so more often because their great weight makes tree-top travelling difficult. Their somewhat protuberant bellies belie the sparseness of their fat and the sinewy strength in their bodies; captive animals can take the husk off a coconut with little effort, and can split hard jungle fruit with consummate ease. Their strength is matched by a great dexterity and a lively curiosity: wild Orangutans have been known to carefully and totally dismantle a camera which was left accidently in the forest; nothing was broken.

Most of the Orangutans' food is fruit, supplemented by leaves, shoots, bark and animal protein from insects, lizards, bird nestlings and eggs – in fact, anything they can get their hands on. They range widely through the forest and have an uncanny ability to locate fruiting trees. It is impossible to see very far in the forest, but Orangutans seem to acquire a knowledge of seasons and memory of the position of flowering trees so that they can predict when different trees will be fruiting. They also watch the movements of flocks of fruit-eating birds such as pigeons or hornbills and use these to locate fruiting trees. Orangutans generally range on their own or as

Orangutans (*Pongo pygmaeus*) are generally solitary animals, except for mothers and their young. Even when Orangutans meet in a large fruiting tree they barely acknowledge one another's presence.

The Agile Gibbon (*Hylobates agilis*) has a peculiar distribution which includes southern and central Sumatra and south-west Kalimantan, suggesting that these areas were once joined when sea levels were lower and Sumatra, Borneo, Java and the Malay Peninsula were all one land mass.

mother-child units, but several may be seen together in a large fruiting tree, generally with very little interaction. Once a good fruit tree has been found, an Orangutan may spend the entire day within it eating, and will rarely range more than a few hundred metres in a day. At the end of the day Orangutans each build a nest of branches and twigs folded down onto a firm branch. They use each nest only once and observations of old nests can be used when assessing the population size of these animals.

Orangutans live as single animals, although the young may stay with their mothers for about six years. Adult males tend to avoid one another, spacing themselves by loud, bellowing 'long calls'. When they do meet, however, they give violent displays and sometimes engage in combat. There is a suggestion that as such encounters increase, levels of courtship, and therefore reproductive rates, decrease. In this way the population density is regulated.

Females are mature at about ten years of age, and will breed every five years or so until they are about thirty years old. Young Orangutans are completely dependent on their mothers for the first couple of years, and generally stay with their mothers for about six years until after the next baby is born. Males and females do not form lasting relationships, but each male tries to cover as many non-pregnant females as possible. The males are not involved in caring for the young at any stage.

Gibbons

Gibbons are the smallest apes and the ones least closely related to man. They weigh 5–6 kilograms (11–13 pounds) and have long arms and hands which they use to swing through the forest canopy beneath the branches, sometimes launching themselves into space to reach the next tree in a manner fit to disgrace even the greatest Olympic gymnast.

There are nine gibbon species found throughout South-east Asia, with the Agile or Dark-handed Gibbon (*Hylobates agilis*) found in central and southern Sumatra and south-west Kalimantan, the Bornean Gibbon (*H. muelleri*) found over the rest of

Borneo, the White-handed Gibbon (*H. lar*) found in northern Sumatra, and the black Siamang (*H. syndactylus*) found throughout Sumatra. In terms of ecology and behaviour they are all very similar, with the exception of the Siamang which is nearly twice as heavy as the others, has smaller territories, and a diet in which leaves play a larger role.

Gibbons are strictly monogamous and territorial. A gibbon group comprises the adult pair and their children, often an infant and an older sibling. Male Siamangs take complete care of their infants from when they are weaned at about one year old until they gain independence in the third year, but the males of the smaller gibbons have relatively little to do with the upbringing of their offspring. Three to four gibbon groups may live within a square kilometre, with some overlap between neighbouring groups. About two-thirds of a group's range is unassailable space which is vigorously defended against attack. This territory is often defined by physical features such as ridge tops or especially large trees from which gibbon calls can carry a long way, and from which their singing displays can be easily seen.

A visitor to the forest will probably first become aware of gibbons through their haunting songs. The males sing around dawn, and later in the morning duets are given by the adult pair. Apart from the territorial purposes of singing, it seems that this activity, which requires the co-ordination and co-operation of both adults, maintains and reinforces the pair bond. While all gibbons are able to sing, not all sing as well as one another; some individuals can be recognized by their 'croaky' voices or inexpert notes. Young animals can also be heard practising with their parents.

Like young Orangutans, young gibbons look as though they should make good pets. They are certainly endearing, but as adults they can be vicious. By this stage, however, the damage is done and the animal is generally destined to live out its days in a restrictive cage without contact from humans or other gibbons. It is interesting that many forest cultures forbid or avoid the killing of gibbons; maybe this is because of their ghostly songs or because of their remarkably human looks. This is good for the gibbons, but it cannot save them from the relentless advance of forest loss which affects all the species.

Proboscis Monkeys

The pendulous-nosed, full-bellied, reddish-brown-faced Proboscis Monkey (*Nasalis larvatus*) is a large, handsome monkey with adult males weighing in at 20–24 kilograms (44–54 pounds). It was described in one of the earliest reports, in 1848, by Hugh Low as follows: 'It is remarkable for its very long nose; it is a very fine monkey, in size approaching the orangutan, but much less disgusting in appearance. It is furnished with a very long tail, and its fur is particularly fine, and of a pretty fawn colour; its head is small; it feeds on fruits'. Not everyone held the animal in such high regard; the famous British Colonial Officer, Charles Hose, opined that 'his appearance is highly ludicrous, he rejoices in a pendulous fleshy nose which droops at the end almost over his mouth. This appendage has no apparent use, and is not even decorative'. It is sometimes known derogatorily as the Dutch Monkey among a wide number of other local names. The female is petite in comparison, being only half the weight of her mate, and her nose begins to droop a little only when she becomes senior in age.

Many arguments for the function of the large nose have been advanced, from the possibility that it helps the monkeys to breathe when swimming underwater, to its use as a resonating

The pendulous-nosed Proboscis Monkey (*Nasalis larvatus*) lives only in Borneo and although it is most common in mangrove forests, it can also be found far inland. It feeds almost exclusively on leaves and has a complex stomach to aid its digestion. Proboscis Monkeys live in large groups comprising an adult male and his harem of females with their offspring.

chamber when calling or as an aid to losing heat in the hot coastal forests this monkey frequently inhabits. While attractive, none of these theories holds up under close examination, and it may be simply that females prefer males with larger noses and so this female selection leads to genes for large noses to persist within the population. The adult males compete vigorously for mates, with a typical group comprising the adult male and three or four females, although larger groups have also been observed. Some of the larger groups are all males, because the excess males have to form bachelor groups, often led by a large sub-adult. Unlike most monkeys, they are not fiercely territorial and it is quite common for groups that meet during the day to mingle and even sleep in the same trees.

Proboscis Monkeys are renowned for their swimming ability and aquatic feats; they have been seen to swim underwater for nearly half a minute. Their hind feet are partially webbed and this must help both with swimming and when they walk on the soft mud of the mangrove forests. The rivers they cross with alacrity are also inhabited by Estuarine Crocodiles (*Crocodylus porosus*) and these must represent serious potential predators. It is therefore interesting that when these monkeys swim they do so with hardly any noise or splashing which might attract a crocodile's attention.

The Proboscis Monkey is primarily an inhabitant of undisturbed, tall coastal swamp and riverine forests, although it is not found in all such areas. Its almost total absence from inland forests may be associated with the generally poor soils on which these forests grow. Consequently the quality and quantity of food produced in these forests cannot match that provided by the coastal and riverine forests growing on richer alluvial soils.

Most monkeys are primarily herbivorous but will take animal food when the opportunity arises. Proboscis Monkeys, however, appear to be totally herbivorous, not even being tempted by the rather easy pickings of the shoreline. Somewhat less than half the diet comprises leaves (mainly the soft and more protein-rich young leaves), the remainder being made up of fruit. The Proboscis Monkey's stomach is highly efficient at fermenting and digesting the leaves it takes, so efficient in fact that it has to avoid 'easy' food such as soft fleshy fruits. The sugars in these would be digested so quickly that there would be a painful, bloating build-up of gas in the stomach which could kill the animal. Thus Proboscis Monkeys select dry, bitter fruit which are digested more slowly. Digestive efficiency is probably also the reason they avoid animal protein – excessive quantities of protein would produce a high concentration of ammonia in the stomach which would make the animals ill.

Leaf Monkeys

The graceful forms of leaf monkeys are commonly seen in the forests of Sumatra and Kalimantan. They may be sighted streaming in a graceful arc from one tree to the next, or spotted quietly resting and basking high in an emergent tree, betrayed by their long, elegant tails hanging down from the branches above. Sumatra has four species of leaf monkey on its mainland and eastern islands. In addition, a remarkable pair of species is found on the Mentawai Islands to the west. The most widespread Sumatran species in inland forests is the Banded Leaf Monkey (*Trachypitheus melalophos*) with eleven subspecies all with different patterns and colours. Some are very poorly known, perhaps from a handful of museum skins, such as the white, ghostly subspecies from northern Lampung and southern South Sumatra, or the very dark subspecies from eastern North Sumatra. In the very different habitats of coastal, riverine, swamp and mangrove forests is found the relatively common

The distinctively marked Thomas' Leaf Monkey (*Presbytis thomasi*) is found in the extreme north of Sumatra, mainly in Aceh Province. This elegant monkey can easily be seen in the accessible parts of Mount Leuser National Park near Bohorok or Ketambe in the Alas Valley.

Silvered Leaf Monkey (*Trachypithecus auratus*) which has metallic grey fur and a dark face. The infants are a rather remarkable bright orange. Kalimantan has five species of leaf monkeys including three not found outside Borneo, the Grey (*Presbytis hosei*), the Maroon (*P. rubicunda*) and the White-fronted (*P. frontata*).

All these species have many ecological features in common: they are all largely arboreal, feeding on fruit, leaves and shoots as well as small animals if the opportunity arises. They have large stomachs divided into a number of sacs, which allow the slow passage of a large quantity of leaves; these are fermented in the upper part of the stomach and broken down into digestible compounds – a similar process takes place in cows and deer. At the same time, toxic compounds in the leaves are neutralized. The large mass of leaves in the stomachs of a leaf monkey can constitute as much as a quarter of its body weight.

Leaf monkeys typically form groups of five to eight individuals with a single adult male and two or more adult females and their offspring. The mothers tolerate the abduction of their babies by the other females, which may even suckle them but without the special care of their natural mothers. The leaf monkey groups seem sedate and serious, they are almost exclusively arboreal, eating a relatively limited range of foods, and do not need to vocalize to co-ordinate group activities as they would if they were foraging at all levels of the forest.

Nocturnal Gliders, Leapers and Those that Potter

Flying squirrels may be seen by the patient observer around dawn and dusk when they are most active. They glide between trees by stretching out their legs to tighten the folds of loose skin along each side of their bodies. There are two groups of flying squirrel species, the large and the small. The former are found in the upper canopy, where they glide between the trunks of giant trees and consume fruit and young leaves. Logging disrupts their lives, but they are still able to move with ease around much of the logged area and indeed can often best be observed in such places. Unlike many nocturnal animals, they travel through the forest in small groups. The smaller flying squirrels are most frequently found at lower levels of the forest where they eat a richer diet including insects and tree sap. They are surprisingly manoeuvrable in 'flight' and can steer a winding path between the trees.

The Colugo or Flying Lemur (*Cynocephalus variegatus*) is a peculiar animal, classified within its own taxonomic order, looking and behaving like a cross between a flying squirrel and a sloth. It is rarely seen since it is largely nocturnal, well camouflaged and an inhabitant of the upper canopy. Like a flying squirrel it has a large flap of skin between its hands and feet, but this also extends to its tail and, unlike the flying squirrels, it does not have a hard cartilaginous spur on its wrist to help 'stow' its wing away neatly. As a result it is rather more clumsy than the squirrels in its four-footed movements, but rather more efficient as a glider. In fact, it prefers to cling onto tree trunks and hang from large branches. It eats mainly leaves and some flowers, a diet more typical of larger animals, and it appears to compensate for this by spending long periods inactive, like a sloth.

At dusk one may be lucky to see small shapes leaping between the stems of saplings and small trees. These are the tiny Tarsiers (*Tarsius bancanus*), primates that could easily sit in the palm of one's hand. They are immediately recognizable by their long, thin tails and enormous eyes. These eyes make them efficient nocturnal predators, able to pounce on insects and other invertebrates, as well as lizards, frogs and even birds. Tarsiers live in small family groups which centre their activity around a hole in a tree. Females give birth once a year to a large

The Slow Loris (*Nycticebus coucang*) is a small nocturnal primate that feeds on large insects, nestling birds, lizards and frogs. Lorises are sometimes caught and offered for sale, but their appealing faces and fluffy fur belie the fact that they have very sharp teeth and are fully prepared to use them. They climb very deliberately, rather like a chameleon, rarely jumping. When approaching living prey, they move cautiously, extending an arm slowly until they can grab the prey.

ABOVE Some species of bat such as this Schreiber's Long-fingered Bat (*Miniopterus schreibersii*) form nurseries deep within caves where the young are left while the adults go out to feed. Amazingly, the nursing mothers are able to locate their own offspring among the hundreds or thousands of naked young, presumably through an acute sense of smell and perhaps of hearing.

BELOW One of the beautiful large caves with extravagant natural architecture near Lho'Nga in western Aceh, Sumatra. Caves are full of interest because the animals within them are specially adapted to the dark and very humid conditions.

single young which is nearly a quarter of its mother's weight at birth.

The Slow Loris (*Nycticebus coucang*) is a nocturnal, solitary creeper which prowls around the lower and middle canopies searching for insects and fleshy fruits both by sight and by using its good sense of smell. It has a short tail, and climbs very deliberately, rather like a chameleon, rarely jumping. When approaching living prey, it moves cautiously, extending its arm slowly until it can make a quick grab. Like the Tarsier, the adult female gives birth to a single young each year. This is carried constantly for a few days, but then 'parked' on a small branch while the mother goes hunting. Both tarsiers and lorises can be found at night by shining a torch from about eye level into the forest and looking for the reflections of red light from their eyes. Both will generally freeze when disturbed, but the loris is able to outstare even the most persistent observer, remaining perfectly motionless for hours on end.

Bats and Caves

One of the great pleasures and surprises for a visitor from temperate areas is the wonderful filling of urban, suburban and forest skies at dusk by bats. Even in an average town, some twelve species can be seen at a single glance, although it takes practice to distinguish them. Kalimantan alone has about ninety species of bats, some of them very little known. The majority of the species are relatively small and insectivorous, with faces designed to aid echo-location and not to appeal to humans. But the larger species are frugivorous and have much more sympathetic faces, not unlike foxes and certain breeds of dogs. Some species live among palm leaves or in tree holes, but many more live in caves, generally in limestone areas. Examples of

these are the insectivorous Wrinkle-lipped Bat (*Tadarida plicata*), which lives in enormous roosts numbering hundreds of thousands, and the Cave Fruit Bat (*Eonycteris spelaea*). The significance of these small animals is often overlooked, but they play important roles in controlling insect populations, and a few species such as the Cave Fruit Bat pollinate commercial fruit trees, particularly the fabled durian, and thus have a quantifiable economic value.

Blind Dolphins

In the mighty Mahakam River in East Kalimantan lives one of Indonesia's largest and most secretive mammals, the Pesut or Freshwater Dolphin (*Orcaella brevirostris*) which grows to about 2 metres (6 feet) long and 100 kilograms (220 pounds) in weight. This is the same species that inhabits estuaries and inshore waters from India to northern Australia. It also enters fresh water in large rivers such as the Irrawaddy, Mekong and the Mahakam, but does not occur in all seemingly appropriate rivers. It has a rounded head rather than the bottle-nosed shape of the familiar oceanarium species, and is a pale blue-grey colour with a lighter grey belly.

The Pesut or Freshwater Dolphin (*Orcaella brevirostris*) is found in the turbid waters of the Mahakam River in East Kalimantan. It has little need of eyes in such conditions, but it does rely heavily on sonar to find its way and to catch fish. This is the same species that inhabits estuaries and inshore waters from India to northern Australia.

Unlike marine dolphins, which have well-developed vision and live in quite clear, deep waters where fish are widely dispersed, the Pesut inhabits shallow, murky waters where fish are relatively abundant. It has little need of eyes but it does rely heavily on sonar. The sonar frequencies it uses are much higher than those used by marine species, and it is not thought to often swim out to sea.

The Pesut has never been the target of traditional hunting in Indonesia, but it was felt to be in need of legal protection in 1975. As with many dolphin species, conflicts with fishermen were increasing. The Mahakam lakes area around Lake Semayang, where the Pesut seems to be most common, is very rich in fish and has attracted many immigrants who fish without restraint, even setting nets right across the major tributaries at times. The nets are set well below the water surface to allow boat traffic to pass unhindered. The Pesut are inadvertently caught and either drown or break through the nets and may then be considered so destructive that they are killed as pests.

Birds

There are 465 bird species resident in Sumatra, and this is the largest number in the region after New Guinea, reflecting its size, habitat diversity, and proximity to the Asian mainland. In contrast with some other islands, only thirteen species are endemic including the Red-billed Partridge (*Arborophila rubirostris*), Salvadori's Pheasant (*Lophura inornata*), Bronze-tailed Peacock Pheasant (*Polyplectron chalcurum*), and the Shiny Whistling Thrush (*Myiophoneus melanurus*) all of which appear to be restricted to montane forests. Kalimantan has nearly 450 resident species and two-thirds of the total species list are found on both islands. The island of Borneo has thirty endemic species but it is possible that ten of these are confined to the high mountains of Sabah and Sarawak. Of the endemics found in Kalimantan, notable species include the small White-fronted Falconet (*Microheirax latifrons*), Red-breasted Partridge (*Arborophila hyperythra*), the impressive Bulwer's Pheasant (*Lophura bulweri*) the male of which has a large white tail, bright blue wattles hanging from its cheeks and blue 'horns' above its head, and the exquisite ground-dwelling Blue-banded and Blue-headed Pittas (*Pitta arquata* and *P. baudi*).

Interestingly, the upper montane forests have very few species of birds and most of these are also found at lower altitudes: in northern Sumatra there are just nine species, eight of which are also found lower down. Higher up in the dwarf subalpine zone, however, a few more species are found and they are largely restricted to this zone. Characteristic and conspicuous birds of the highest mountain tops are the Sunda Whistling Thrush (*Myiophoneus glaucinus*), Scaly Thrush (*Zoothera dauma*), and the Island Thrush (*Turdus poliocephalus*). It is a strange feeling to be in an area surrounded by tropical rain forest and yet hear such very thrush-like songs around dawn and dusk. In many cases these birds are quite tame, presumably because they see humans so rarely.

In the forest around dawn, there is a veritable 'rush-hour' for the birds and it is possible to see and hear a large number of beautiful species. In lowland forest these will include the barbets with green bodies and multi-coloured heads, the shy red-and-black trogons, the rich-green broadbill and leafbirds, the black and neon-blue fairy bluebirds and the yellow-and-black orioles. Among the most evocative is the Argus Pheasant whose 'ki-au' call regularly frustrates those who try to catch a glimpse of it, and the hornbills whose noisy flight and raucous calls soon become familiar.

Argus Pheasants

Walking through remote areas of forest in Sumatra and Kalimantan, one may be startled to come across a roughly circular area several metres across which has clearly been swept clean of leaves and small plants. The perpetrator is the male Great Argus Pheasant (*Argusianus argus*), one of the largest birds in the forest. Its call and its large moulted feathers are very familiar to those who have spent time walking in these areas, but it is excruciatingly difficult to see the bird itself. Both male and female live more or less solitary lives throughout the year except for the short breeding period. The male has subtly patterned wings and long tail feathers, and he shows these off to the rather dull females in a beautiful dance in the depths of the forest on an open, level stage of bare earth which he scratches clear of small plants and dead leaves. One observer has noted that dancing grounds always have an 'exit' into dense forest through which the male could escape should he be disturbed. The dance comprises three phases, beginning with a

LEFT The 'ki-au' call given by the Argus Pheasant (*Argusianus argus*) is one of the most characteristic sounds of the forests in Sumatra and Kalimantan. While its song may be well known, it is exceedingly difficult to see because it will scuttle through the undergrowth at the slightest hint of danger.

BOTTOM OF PAGE The Rhinoceros Hornbill (*Buceros rhinoceros*) is found in both Sumatra and Kalimantan and is surprisingly common in the remaining forest areas. It is mainly frugivorous but will eat small animals when the opportunity permits.

period of agitation during which the male sweeps his tail to and fro while fussing around the female. He then raises his tail and wings to form a magnificent vertical fan of grey ocellations. Finally, he bows repeatedly to her and mates with her if she has been suitably impressed.

Hornbills

Hornbills are among the largest forest birds with some of the species well over a metre long. They are all fruit eaters although most will also try to catch anything that comes their way such as lizards or even small birds. The 'horn' or casque is generally hollow or spongy, but in one species, the long-tailed Helmeted Hornbill (*Rhinoplax vigil*), it is solid. This 'hornbill ivory' has been traded for hundreds of years and is carved into delicate objects.

The hornbills' nesting behaviour is truly remarkable. A hole in a tree is selected, and a false floor is constructed if necessary by throwing in twigs. After mating, the female imprisons herself within the tree by plastering her droppings around the edge of the hole. The male assists, bringing clay which he pats into place using his large bill. The female remains incarcerated and dependent on the male for food until the chicks have fledged. Once a pair has found a suitable hole it will continue to use it for many years, which is not surprising, given the general scarcity of appropriately sized and positioned holes.

There are eight species of hornbills in Kalimantan and ten on Sumatra. A community of seven species has been studied in detail in East Kalimantan. It is interesting in that although they are all superficially similar, they all have different habits. For example, two of the species are nomadic, flocking species, three form territorial pairs, and two live as territorial communal groups. These differences reflect differences in diet; thus the nomadic species are able to avail themselves of the relatively rare but more energy-rich fruits but only by using more energy to search for them, whereas the hornbills which are limited to a defined territory eat energy-rich fruit only when it is available in their area, and between these periods they rely on the energy-poor fruit (such as figs) and animal prey but they do not have to expend much energy to find them.

Freshwater Fishes

Tiger Barbs, Harlequins, Rasboras, Glass Catfish, Black and Marbled Lancers, Silver Sharks, Scissortails, Coolie Loaches, Clown Loaches, Spiny Eels, Archer Fish, Flying Foxes and the expensive Asian Bonytongue are just some of the popular aquarium fishes which originate from Sumatra and Kalimantan. None of these is particularly large, but the larger rivers harbour giants, such as the Giant Sucking Catfish (*Bagarius yarrelli*) which grows to more than 2 metres (6 feet) long. Downstream in the estuaries the world's largest eel, *Thyrsoidea macrurus*, is sometimes caught, measuring over 3 metres (10 feet), in roughly the same areas as some of the world's smallest vertebrates, such as the tiny yellow and black goby *Pandaka pygmaea* which grows to just 11 millimetres (less than half an inch). In the early-morning fish-markets in towns along the larger rivers about seventy species can be seen in a single morning, though this is

A fisherman checking his gill net on Lake Semayang in East Kalimantan. The abundant fish catch from this and adjacent lakes supplies a staggering one-third of all the fish eaten on the densely populated island of Java. There are indications, however, that overfishing is reducing the success of the fisheries.

not the case in Medan, for example, where the nearby rivers have become polluted with industrial and plantation wastes and chemicals.

At the moment, however, Sumatra and Borneo are known to have 272 and 394 species of fishes respectively in their rivers, estuaries and lakes, about 85 per cent of which are completely restricted to fresh water. Many species are common to both islands and, as is the pattern in other groups of animals and plants, Borneo has more endemic species: 149 against 30. New species of fishes are still there for the finding, particularly in the blackwater rivers of eastern Sumatra and southern Kalimantan and in the headwaters, but forest loss and its various consequences and unsustainable exploitation of various types are taking their toll, perhaps extinguishing species before they are even discovered.

Asian Bonytongues

The Asian Bonytongue or Arowana (*Scleropages formosus*) is the most highly prized aquarium fish in the world, with the price for a good specimen reaching thousands of pounds. It has a straight back with the dorsal fin quite close to the tail, the scales are large, and there are two short barbels projecting from the lower lip of the large and sloping mouth. Asian Bonytongues are mouth brooders, that is, the adults take the eggs into their mouths to develop and hatch. The fry continue to shelter inside until they are about 6 centimetres (2 inches) long and their yolk sacs are fully absorbed. Although Asian Bonytongues skulk near the river or lake bottom during the day, they rise to the surface at night to feed on frogs, small fishes and insects which fall onto the water surface.

Interestingly, different populations have different colour fins – green, yellow, white, red – but the variety most sought after is the golden-red, which fairly glows with shimmering fire. The white-, yellow- and green-finned varieties have a relatively low commercial value, and their wild populations are probably not under threat, but the numbers of the red and golden-red varieties from one area of the great Kapuas River need some

specific management control. All international trade in the species was banned in 1975, and the Asian Bonytongue was formally protected in Indonesia in 1980. The hope is that in due course captive breeding will provide good quality specimens at prices which undercut the wild-caught fish. Although there have been claims of captive breeding for many years, until recently most of them have in fact been fronts for poaching operations. There is now international agreement that Indonesia should enjoy an annual export quota provided that an increasing proportion of these come from proven captive-bred stocks.

Conservation

Both Sumatra and Kalimantan still have considerable areas of forest remaining; the latest figures indicate 49 per cent and 75 per cent respectively. In Sumatra these are in relatively small blocks, whereas in Kalimantan attrition has come from the coast and most forest areas are contiguous. Much of this forest has already been logged, however, and the area of relatively pristine forest is probably rather small. Logged forest looks a real mess from the logging roads, but when considered in the context of an entire concession, such disturbed areas support virtually identical wildlife communities to the pristine areas. Logged forest can and does regenerate – if given the chance. Many forests which have been logged by the concession holders or their contractors more or less according to the rules and regulations of the Forestry Department are subsequently subjected to further abuse by a succession of small-scale illegal loggers who open the way to farmers, who clear the remaining trees themselves in order to grow crops.

Statistics demonstrate that the area of loggable forest that should be sustaining the economically important timber industry is diminishing year by year. The government is very concerned about this but even the new moves to stem the erosion of renewable wealth are unlikely to take effect before much more has been lost.

Java and Bali

'As the sun rose, bluish mountains came up from the sea, grew in height, outlined themselves, and then stood out, detached volcanic peaks of most lovely lines, against the purest, pale-blue sky; soft clouds floated up and clung to the summits; the blue and green at the water's edge resolved itself into groves and lines of palms; and over sea and sky and the wonderland before us was all the dewy freshness of dawn in Eden.'

So Eliza Ruhamah Scidmore wrote of her first view of Java towards the end of the last century in her vivid travel book *Java: The Garden of the East*.

The name Java conjures up different images for different people. For many it speaks of the frantic business of the capital city, Jakarta, where modern skyscrapers jostle with old colonial buildings and modest single-storey houses, where a complex system of overpasses and toll roads operates alongside the old road system crammed with buses, cars and motorbikes. For some people Java speaks of the ancient civilizations of Central Java where the famous eighth-century Buddhist temple of Borobudur can be seen and where, in more recent times, the ladies of the court practised the art of batik making in which designs are painted in wax before being dyed; the execution of this beautiful art form can still be seen, particularly around Yogyakarta and Solo, and the finished products are exported all over the world. Others recall that Java is the most densely populated island in the world.

ABOVE The south coast of Java is rough and steep, and there are many areas, such as shown here in Meru Betiri National Park, where cliffs plummet straight into the often turbulent sea. Not far to the east is some of the world's best surfing.

BELOW Jalan Sudirman in Jakarta, sometimes known as 'Money Mile', is one of the busiest and most developed streets in the nation's capital.

Java can have all these connotations, but also that of so much beautiful rural scenery and even forest. The combination of a fertile volcanic soil and industrious rural population have resulted in finely terraced hillsides in shades of emerald green and brownish-yellow from the young and mature rice plants, with some areas harvesting three crops of rice a year. Most of the remaining wild areas are unsuitable for agriculture, being montane, and some are difficult to reach because they tend to be off the beaten track; others, such as the Baluran and Gede-Pangrango National Parks, are easily accessible from the main roads. Bali, likewise, is best known for one or two facets, in this case its beaches and colourful Hindu religious practices, but in west Bali a large area of natural rain forest remains.

Both the flora and fauna of Java and Bali are rather poor in species, but the islands' relative isolation from Sumatra, Borneo and Malaysia has resulted in more endemic species than might be expected – for example, 7 per cent of the resident birds and 12 per cent of the mammals. The majority of these are confined to western Java and western Bali in the remaining lowland rain forests.

Forest and Savannah

In general, the further east and north one travels in Java and Bali, the more seasonal is the climate and this is clearly reflected in the vegetation. For example, the remaining forests in the south and west of Java have luxuriant tropical rain forest very similar to that found in southern Sumatra, although mature forest is found only on the largely inactive volcanoes.

ABOVE A tranquil rural scene at the edge of Mount Halimun Reserve is typical of many areas in West Java. Bird and fish life are scarce in these areas because of the abuse of pesticides, but new controls may, in the long term, result in some improvements.

RIGHT Rain forest in Ujung Kulon National Park at the western tip of Java. The famous eruption of nearby Krakatau Volcano in 1883 damaged this region and it is possible that these two 'pollarded' trees may have had their tops knocked off when huge tidal waves swept through the area.

Most of the original lowland forest in the drier areas of Java (northern coast and eastern Java), and most of the forests on limestone, have been lost to different forms of agriculture and teak plantations respectively, and mangroves and freshwater swamp remain in only tiny disturbed pockets, though even these have retained some of their ornithological interest. There is very little forest left in the lowlands of eastern Java, because the natural deciduous monsoon forest has been subjected to repeated human-induced fire. Baluran National Park is the only place where this type of vegetation can now be seen in any quantity, but even here it is set in a mosaic of fire-climax grasslands.

On Bali, however, forest can still be seen in a major block in the west of the island within the Bali Barat National Park, as well as around the mountains in the centre and east. In the west and parts of the north of the park there is a dry, almost monsoonal, forest and true rain forest occurs only on the tops of some of the mountains. The Balinese are not great mountain trekkers, and there are virtually no paths through the forest. As a result there are some virtually untouched wild areas in the centre of this block.

Volcanoes

Java and Bali have more active volcanoes than anywhere else in the world, with some forty regarded as fully active and many more dormant but still bubbling or providing hot springs for rheumatic pilgrims. Volcanoes are among the most popular destinations for foreign and domestic ecotourists alike, although the bare and relatively recent crater landscapes of Bromo, Semeru and Ijen are not particularly rewarding for those seeking wildlife. Volcanic rock and soil are generally dry, sterile, acid, lacking in organic matter and often hot from the gases below. The vents of sulphurous and other toxic gases preclude the growth of most plants, but there are a few very hardy species which do survive in these Inferno-like surroundings, such as blue-green algae in the sulphurous pools, prostrate bilberries (*Vaccinium*), the red-flowered Blunt Rhododendron (*Rhododendron retusum*), and the simple-fronded fern *Selliguea feei*. The acute slopes are unstable and these hardy plants have long, deep roots to prevent them being swept away. Away from the craters, the fascinating and often beautiful plants of Java's volcanoes can be admired and studied with relative ease, though many are under threat from the encroachment of vegetable farmers, even in national parks.

Where frequent volcanic eruptions burn or smother the trees with ash, the successional change from one species group to another keeps beginning again from scratch. There are plant specialists, with the first colonizers being known as pioneers or nomads. These are fecund, fast-growing, often short-lived species, and if there is no further disturbance they will give way to a second generation of plants tolerant of the shadier conditions. These in turn give way to another group of relatively slow-growing, long-lived species which demand shade.

An interesting long-lived pioneer is the Cemara (*Casuarina junghuhniana*) found eastwards from Mount Lawu on the border of East and Central Java. It can reach 45 metres (150 feet) in height and looks superficially like a conifer but is no relative. It grows as more or less the only tree over wide areas at altitudes over 1,400 metres (4,600 feet), although a variety of grasses and colourful herbs grow beneath it, and a host of epiphytes such as ferns, orchids and the pale-green garlands of the lichen *Usnea* grow among its branches. Its seeds can germinate only in light conditions and when in contact with mineral soil or ash – that is,

A sulphur collector climbing out from the suffocating crater of Ijen Volcano in East Java prior to taking his load to the collecting point at the foot of the mountain. He will make this journey several times each day. The sulphur is used in the manufacture of matches.

where there has been recent volcanic activity, and not in established Cemara forest. The trees drop many fine needles and branches, and these and the grass make excellent tinder, so fires (generally started by man) are common in the dry season, killing off most plants but leaving the thick-barked trees more or less unscathed. Repeated fires do take their toll, however, but the trees are very efficient at regenerating from burned stumps and hidden roots. Cemara can even stand being buried by volcanic ash. Interestingly, if the Cemara forests are protected from fire, then the stand may grow to become large and beautiful, but is doomed to be substituted by a mixed forest of oaks and laurels. New Cemara forests grow up where fire or landslips, lava flows or ash fields provide the right conditions.

A map of natural forest on Java illustrates that in most cases the remaining patches are around the highest peaks, separated from one another by a sea of inhospitable agricultural and urban habitats, intraversable by all but the hardiest and most catholic animal and plant species. Some plants produce pollen and seeds which are adapted for wind dispersal, but many of the shrubs and small trees in the upper montane and subalpine forests produce berries. These represent an important food source for birds and mammals and germinate where the bird or mammal deposits them after their passage through the gut. However, as it takes only about an hour to pass through a bird gut, there is little chance of a seed being taken more than a relatively short distance before being deposited. In fact most of the high mountain birds are rather sedentary and so do not fly between peaks dispersing seeds. Most montane animal species are found only on mountain tops and cannot cross between mountains. These include the attractive red-and-black Warty Toad (*Cacophryne cruentata*) known only from Mount Gede National Park between Bogor and Bandung in West Java, and the Volcano Mouse (*Mus crociduroides*) known only from Mount Gede and Mount Kerinci in southern Sumatra. Many of the beautiful alpine flowers that grace the mountain meadows are in the same predicament; for example, the endemic Busy Lizzie (*Impatiens radicans*) and the rare, endemic red-flowered orchid, *Dendrobium jacobsonii*, from East Java.

Edelweiss of the Spirit World

Up on the highest slopes of the volcanoes one can find groups of shrubs with white furry leaves and white-and-yellow flowers reminiscent of the Edelweiss of the European alps. Known in English as the Javanese Edelweiss (*Anaphalis javanica*), it is in fact found from central Sumatra and West Java through to Lombok in the Lesser Sundas, as well as in Sulawesi. The flowers are generally seen between April and August and are visited by many bees, flies and butterflies on sunny days.

On most mountains one can now find only small specimens up to a metre (3 feet) tall, but this giant daisy is, in fact, capable of growing up to four or even eight times this height on the infertile, volcanic screes which it favours. The largest specimens in Java are apparently in sheltered spots on Mount Sumbing where the 8-metre (25-foot) tall specimens have rough, fissured trunks 15 centimetres (6 inches) thick; they may be over a hundred years old. The most impressive mass of them, however, is in the field behind the crater of Mount Gede which can be visited easily, depending how wobbly your legs feel after the ascent of the mountain – the climb back up Gede from the field can be pretty devastating if you are feeling tired.

On Mount Gede and Mount Agung, Bali's mightiest peak, pilgrim climbers regard this beautiful plant as a gift from heaven and it is their custom to take a fragment down with them as a divine blessing. As one might guess, this habit has not greatly served the needs of the Edelweiss, but enterprising villagers on the lower slopes of Mount Agung near Besakih cultivate the plant and offer pieces for sale, probably for divine profit.

Mammals

Java has seventeen endemic species of mammals: six bats, one leaf monkey, one gibbon, one squirrel, six rats, one deer and one pig. A further five species (three bats, a hare and the leopard) are not known from anywhere else in Indonesia although present elsewhere in Asia. Four of the larger and rarer species are described below.

Javan Rhinoceros

The Javan Rhinoceros (*Rhinoceros sondaicus*) is the largest animal in Java with a shoulder height of over 1.5 metres (nearly 5 feet), a body length of 3 metres (10 feet), and a weight of up to two tons. It has three distinct folds of skin, one around the neck, one over the shoulder blades and one around the hips. Its upper lip is long and prehensile and it uses this to pull leaves and twigs towards its mouth when browsing. All males have a horn, visible even in a new-born calf, but most females have only the suspicion of a bump. Even in males the horn never grows very long, with the record being a meagre 25 centimetres (10 inches) and the average a mere 15 centimetres (6 inches).

Javan rhinos enjoy wallowing in mud and their habitat is well supplied with suitable wet places. They prefer low-lying areas, but there are records from the last century of rhino tracks extending all the way up Mount Gede-Pangrango to just over 3,000 metres (9,800 feet). They tend to live largely solitary lives, forming adult pairs only temporarily at breeding time and cow–calf pairs for the three or four years that a mother looks after its young. They are browsing animals, eating young leaves, shoots, twigs and fallen fruit along the forest margins.

About 150 food-plant species have been identified, most of these being typical of secondary growth. Interestingly, the damage a rhino causes to a plant when it feeds is not enough to kill it, and the plant generally responds by growing new shoots at precisely the height a rhino finds convenient for feeding. The rhino is, therefore, a positive influence on its habitat for its own needs, and the paths it regularly travels take on a distinct and unique appearance.

This rhinoceros is not very aptly named. Until recently it was thought to survive only at the western tip of Java in Ujung Kulon National Park but a hundred years ago it was found from the Sundarbans of Bangladesh and northern Vietnam (where a small population persists) through to Sumatra and Java. In prehistoric times it may even have lived on Borneo but, if so, it has been extinct there for a long time. Today the main population is in Ujung Kulon where some sixty animals remain in an apparently stable population. Some visitors to the park may catch a glimpse of a large grey behind or hear a snort and crashing of branches in the undergrowth, but the majority are not so lucky. The rhino used to occur throughout Java and favoured disturbed vegetation, which brought it into contact with people. It was apparently a belligerent beast, and when people obtained firearms they started to get their own back.

Hunting used to be a major threat to the Javan Rhinoceros, but with so few animals remaining it is no longer the rhino's biggest problem. Perhaps the greatest present threat is its own rarity and confined distribution. A few years ago a disease similar to anthrax caused the death of a number of animals and it is possible that some disease could wipe out the entire population, or at least such a large proportion that it would be unable to recover. In the light of this, a major captive breeding programme has been proposed. However, the lack of appropriate biological information, the large numbers that would have to be removed from the park, the lack of any assurance of success, and the proven ability of the animals to withstand both disease and the enormous Krakatau eruption one hundred years ago, all argue strongly against such a programme.

Banteng

The second largest wild animal in Java and Bali is the Banteng (*Bos javanicus*). Large handsome cattle, standing 1.5 metres (nearly 5 feet) at the shoulder and weighing up to 800 kilograms (1,760 pounds), male and female Banteng are very distinct: the male is a dark chestnut brown and the female mid-brown, but both have a contrasting white rump-patch and stockings. Distributed from Burma to Borneo and Bali, Banteng are missing from Malaysia and Sumatra although they did once occur in Malaysia. They are probably the ancestors of the domestic Asian cattle and in some areas where both are found it can be difficult, even for the cattle owners, to tell young wild and domestic animals apart at a distance. The adult male wild Banteng is highly conspicuous, however, by his colour and the high ridge along his back.

Banteng are not beasts of dense primary forest, preferring instead rather more open areas such as clearings and river banks, where they graze on grasses. They generally live in herds of twenty-five or more animals containing one adult male; the surplus males group together to form bachelor herds. Herding is to their advantage because animals in a group are less likely to fall prey to leopards (and, in the past, to tigers). The herd, as in the case of elephants, is generally led by an old cow.

Wild Banteng are largely nocturnal but this may be a reaction to human hunting pressure. When guns became available, the wild populations suffered severely and they have been exter-

ABOVE Banteng (*Bos javanicus*), such as these in Baluran National Park, are the second largest animals on Java. They are not beasts of dense primary forest, preferring instead rather more open areas such as clearings and river banks. They live in herds of up to twenty-five animals which will contain one adult male but be led by an old cow.

RIGHT The Warty Pig (*Sus verrucosus*) is endemic to Java, and is distinguished from the common Wild Boar by the three pairs of warts on the face of the adult male: at the corner of the jaws, below the eyes and on the snout. The status of the pig recently caused some concern, but it does not seem to be as rare as was feared.

minated from many areas but are still found in the larger patches of Javan lowland forest such as Ujung Kulon, Alas Purwo and Baluran. Banteng are now regarded as a vulnerable species. They seem able to adapt to living in logged forest, not least because many of their food plants colonize the open ground, but they are unable to cope with the extremes of habitat change which come with forest clearance. Apart from the obvious threat of habitat destruction, Banteng also face the threat of interbreeding with domestic cattle and hence losing their genetic identity, but this has yet to be investigated in detail. Even so, it is known that the beasts at Pangandaran Reserve, at the south-east corner of West Java, are hybrids, and this may also be the case in other reserves.

Warty Pig

The endemic Warty Pig (*Sus verrucosus*) is similar to and just as variable in size, shape, colour and ecology as the common Wild Boar or Pig (*Sus scrofa*) which is found throughout the temperate and tropical regions. It is distinguished from the common pig, however, by three pairs of warts on the face of the adult male at the corner of the jaws, below the eyes and on the snout. The status of the pig recently caused some concern because it seemed that it was no longer reported from areas where it was once known, and loss of habitat and hybridization with or competition from common pigs were thought to be the cause. In fact, it does not seem to be as rare as was feared and could even be described as common in some areas.

The Warty Pig is not found above 800 metres (2,600 feet) whereas the common pig can be found at all altitudes. Its preferred habitat is extensive areas of lowland secondary vegetation, particularly teak plantations in Central Java where there is a mixture of different-aged trees and grasslands with clumps of bush or heavily disturbed forest. It also frequents coastal forests where it is often the only pig species present. In contrast, the common pig can be found in all habitats from primary forest to agricultural land close to human habitation. Another difference between the two species is that the Warty Pig roams in small groups of four to six, whereas the common

pig can sometimes be seen in aggregations ten times this size.

The only predator of significance faced by the Warty Pig is man. Despite the fact that the vast majority of Java's population adhere to the tenets of Islam and therefore cannot eat pig meat, pigs are a popular hunting quarry. They are hunted for sport with firearms or spears and dogs, and are also snared and illegally poisoned to protect crops, although Warty Pigs raid crops less frequently than the common wild pigs. The meat is then sold in the markets to those whose religion permits them to eat it.

Bawean Deer

The Bawean Deer (*Axis kuhli*) is one of the rarest deer in the world. It is, and always has been, restricted to the small volcanic island of Bawean, 150 kilometres (90 miles) to the north of Java, making it the most restricted deer species in the world. It is quite small, reaching only 70 centimetres (30 inches) at the shoulder, with a short face, white throat, a dark stripe down the middle of its brown back, and a bushy tail. It was recognized as a distinct species back in 1836 when the German explorer Salomon Müller examined a domestic herd kept by the governor of the Dutch East Indies. It is presumed to have descended from a now-extinct deer species which was present in Java some 10,000 years ago when the sea levels were low and a land-bridge linked Bawean and Java. Very little of the island now has natural forest on it, but teak plantations with scrubby undergrowth are acceptable habitat for the deer, as are grassy glades resulting from fire, where the regrowth vegetation is highly palatable. The total population is in the region of 200–400 animals and, thanks to government programmes to raise the awareness of the 75,000 islanders concerning the protected status of the deer, hunting is far less common now than in the past. However, uncontrolled fires and the conversion of the remaining steep forest land to dryland agriculture remain serious threats. It is hoped that those responsible for the production of teak and those charged with protecting wildlife can come to practical management decisions to ensure the survival of this remarkable deer.

Extinctions

A host of mammals are known or thought to have become extinct on Java and Bali in recent decades or centuries, such as the ghostly white and pungently smelly Moonrat (*Echinosorex gymnura*), the Hispid Hare (*Caprolagus hispidus*), Siamang (*Hylobates syndactylus*), Orangutan (*Pongo pygmaeus*), Sun Bear (*Helarctos malayanus*), Clouded Leopard (*Neofelis nebulosa*), Asian Elephant (*Elephas maximus*), Tapir (*Tapirus indicus*) and, most recently, the Bali and Java subspecies of the Tiger (*Panthera tigris balica* and *sondaica*), the smallest of all the tigers.

Birds

Java and Bali are home to 340 resident species of birds, one of which is endemic to Bali alone and eighteen of which are endemic to Java, or Java and Bali. One of these is the red-billed, grey-backed and white-cheeked Java Sparrow (*Padda oryzivora*) which is a popular cage bird and has been introduced to many other islands in Indonesia as well as beyond; it comes as a surprise to see this rare Javan bird eating scraps of food along the streets in Honolulu.

Birds are very popular in Java, though not necessarily in their natural environment. Young urban cowboys are a frequent sight, toting their air rifles over their arms, scanning the tree tops for bulbuls or tailorbirds. By midday a row of limp birds can be seen hanging from their belts, sewn on a piece of string, testimony to a saddening lack of awareness and resulting in a very sparse town-bird fauna. In contrast, caged birds of all kinds are very commonly seen in Java, the cages often being hoisted on tall bamboo poles above houses in the mornings for the birds to sing. Men on bicycles laden with cages can be seen pedalling gingerly along crowded streets to sell their wares on street corners. Hardly any of these doves, bulbuls, starlings, mynas and other birds are bred in captivity, the vast majority being wild-caught in Java and beyond by bands of professional trappers regularly providing some thirty or more species for sale. Some, such as the Straw-headed Bulbul (*Pycnonotus zeylanicus*) and White-rumped Shama (*Copsychus malabaricus*) have been so intensively hunted that they are extinct in the wilds of Java, and any free-flying birds that may be seen are in fact escaped cage birds. Bird-song competitions are popular events, and prices exceeding $1,000 are sometimes paid for winning birds.

Green and Red Junglefowl

The Green Junglefowl (*Gallus varius*) has a beautiful violet and red wattle and comb, and iridescent green plumage. It is restricted to Java and the islands of Flores and Sumba and likes open ground, being frequently seen in teak plantations. It also visits the edges of villages, but does not appear to interbreed with domestic chickens, which were derived initially from the Red Junglefowl (*Gallus gallus*). This is a larger bird with the familiar red wattles and comb of the domestic chicken. It has a wide natural distribution from the Himalayas through Southeast Asia and prefers the dense cover of primary forest. This does not normally bring it into contact with man and his chickens, so there are few opportunities for interbreeding.

Bali Starling

The most famous bird on Java or Bali, in avicultural circles, is undoubtedly the Bali Starling or Rothschild's Myna (*Leucopsar rothschildi*), the only bird endemic to Bali. This is a smart white bird with a long crest, bare blue skin around the eyes and black tips to the wings and tail. Its natural habitat is the relatively dry lowland forests of northern north-western Bali, much of which has been converted to agricultural uses. It roosts communally in tree holes often previously occupied by woodpeckers. Bali Starlings feed in pairs, eating mainly insects but also taking some fruit.

Caged birds are very popular in Indonesia, but some of the birds that are kept, such as the cockatoo in the foreground, are becoming extremely uncommon in the wild because of the trade. Essentially all these birds are wild-caught rather than bred in captivity.

The Bali Starling (*Leucopsar rothschildi*) is the only species of bird endemic to Bali and one of the most critically threatened birds in Indonesia. Only about twenty or thirty are left in the wild, partly because of habitat loss, and partly because of intensive trapping to satisfy the demand for cage birds. A conservation programme is now under way using birds sent from a number of zoos around the world.

Habitat loss made this bird vulnerable to extinction, but it seems that news of its rarity and exotic locale made it extremely attractive to aviary enthusiasts and zoos, with the result that hundreds of the birds were caught 'to save them from extinction'. These misguided moves have led to today's ridiculous situation in which less than fifty truly wild birds are left in Bali, mainly on an island within the Bali Barat National Park, whereas thousands are kept by hobbyists in North America, Europe, Japan and Indonesian urban centres. Logic dictated that some of these captive birds should be brought back to Bali to boost the wild population, and a programme to this end has been instigated. Unfortunately, the birds are still being trapped, and at least one ringed, captive-bred released bird has ended up for sale in a Jakarta bird market.

Extinctions

Travelling into Jakarta along the toll road from Soekarno-Hatta airport, one looks out at fast diminishing areas of brackish-water fish ponds and grassy swamps of the sort that once were inhabited by the Javan Wattled Lapwing (*Vanellus macropterus*), known only from grassy patches in north-west Java. There have been no records of this bird now for many years and it is feared to be extinct. However, if one should appear it would not be difficult to identify – a long-legged bird about 30 centimetres (one foot) tall, it has a brown back, black flight feathers, a black and white banded tail, black head and neck and, most conspicuous of all, fleshy pink and white wattles beneath the eyes.

A number of other endemic Javan birds give cause for concern because they have been seen or heard so rarely in recent years. These include the large, crested Javan Hawk-eagle (*Spizaëtus*

bartelsi) which lives in forest and open wooded areas in hilly and mountainous regions but for which there are only two recent records; the small, dark, freckled Javan Scops Owl (*Otus angelinae*) with long ear-tufts, for which there is a single recent record from Mount Pangrango, the only mountain from which it is known; and the Pigmy Tit (*Psaltria exilis*), a small and rather dull bird but none the less worthy of concern, for which there are only a few recent records. On Bali the most critically threatened bird is the beautiful Bali Starling described above.

Conservation

The very rarity of the remnant forest areas of Java and Bali gives a great significance to each area and all efforts are required now to conserve what remains. There are enough other areas available for industrial or agricultural development to make any further sanctioning of the destruction of natural areas unjustifiable. The burgeoning middle classes are taking an increasing interest in conservation and some areas such as Pangandaran Reserve and Mount Gede-Pangrango National Park in West Java, and Baluran National Park in East Java are suffering from the attention of too many visitors. It would be ironic if, by the time there is strong vocal and active support for conservation, there were virtually nothing left in Java and Bali worth conserving.

Lake Batur, in the centre of the mountains in northern Bali, is a popular tourist destination beneath the active Mount Batur. On the far side of the lake is the village of Trunyan, home of descendants of the Bali Aga, who lived on Bali before the Hindu colonization in 1343.

Sulawesi

The dominant image of wild Sulawesi is its mountains, which, unlike those of Java and Bali, are mostly non-volcanic, having been formed by folding and uplift rather than by catastrophic eruptions. Fully one-fifth of Sulawesi's land area lies above 1,000 metres (3,280 feet) and the plains, such as they are, are limited to a few coastal enclaves in which have grown the major towns and cities. The view from the top of Mount Rantemario, at 3,455 metres (11,335 feet) the tallest mountain on the island, is of seemingly endless blue-hazed ridges in almost all directions and of unnamed peaks holding the promise of new discoveries. The vegetation here in this windy and lonely wilderness is low, partly because of the harshness of the climate (frosts sometimes occur) and partly because of fires set by generations of hunters to ensure that their quarry have less cover in which to hide.

Cliffs and Caves

There are a number of impressive limestone areas in Sulawesi. The best known are those north of Ujung Pandang around Maros and Bantimurung, and those of Torajaland, but there are also extensive areas near Bone, south-east of Lake Poso at the tip of the eastern arm, scattered around in the south-east peninsula, and on the islands of Buton and Muna. The limestone takes different forms in all these areas: the Toraja limestone is high, craggy ridges, near Bone and on Buton and Muna it is conical hills, and the form around Maros is the dramatic vertical tower karst.

It was Bantimurung in the Maros area that so amazed the British naturalist and traveller Alfred Russel Wallace when he visited it in 1857. He described it thus:

> Such gorges, chasms, and precipices as here abound, I have nowhere seen in the Indonesian Archipelago. A sloping surface is scarcely seen anywhere to be found, huge walls and rugged masses of rock terminating all the mountains and inclosing the valleys. In many parts there are vertical and even overhanging precipices five or six hundred feet high, yet completely clothed with a tapestry of vegetation ... These precipices are ... very irregular, broken into holes and fissures, with ledges overhanging the mouth of gloomy caverns; but from each projecting part

The bubbling, steaming crater of Lokon Volcano near Manado, capital of North Sulawesi. This volcano becomes active every few years but has never caused any fatalities. Lokon is one of ten active volcanoes on Sulawesi and its offshore islands which have erupted in historical times.

The sheer, tower-like limestone hills (RIGHT) near Maros in South Sulawesi are an oasis of cool, dark greenery in a sea of pale green rice and grass. The vegetation (LEFT) is specially adapted to the rapid drying which occurs in such habitats. Within the hills are caves, some of the short ones containing prehistoric paintings, and others winding through the rock for as much as 11 kilometres (7 miles).

have descended stalactites, often forming wild gothic tracery over the caves and receding hollows . . .

Maros is becoming known as a major caving area and here is found Indonesia's second longest known cave, Salukan Kalang, measuring about 11 kilometres (7 miles). The caves harbour a rich invertebrate fauna: there are large crickets, ferocious-looking whip-scorpions with huge, hinged grabbing arms, click beetles, ground beetles, scarab beetles and a host of cock-roaches, sometimes so thick that they form continuous carpets on the cave floor. These animals feed primarily on the faeces dropping from the cave ceiling which is inhabited by thousands upon thousands of swiftlets (small echo-locating relatives of common swifts) and both insect-eating and fruit-eating bats. In the water of long caves specialized animals have evolved which have no pigment in their skin, and much reduced eyes, or in some cases no functioning eyes at all. A number of cave-adapted crabs, shrimps and fish new to science have been found in the Maros area in recent years.

Types of Forest

The varied geology of Sulawesi is reflected in the number of forest types that can be seen: forests on limestone, on infertile ultrabasic rocks, on fertile volcanic rocks, and zones of different forests up the mountains. The lowland forests are best known for a small number of valuable tree species, such as the ebonies, particularly the blackest and hardest species, *Diospyros celebica*, found in northern and central regions. This is greatly sought after by woodcarvers in Bali and thereafter by tourists and collectors. Ebony is being poached mercilessly from reserves and a single raid by Forestry Department officials in 1986 confiscated nearly 700 lengths representing between 200 and 300 trees.

Forests on Limestone

The forests growing on limestone hills differ from those in other lowland areas. Large trees are unable to grow on the steep slopes, which are free-draining and thus dry, and the high concentrations of calcium and magnesium are also important factors. Trees are smaller and represented by fewer species, but the herb flora can be quite rich and contains species not found outside limestone habitats. The response of some plants to the occasional severe water stress is to lose most of their water, appearing completely withered, though retaining the water within the protoplasm of their living cells. When water becomes available again, they revive and appear none the worse for wear. Other plants on these hills, such as the *Kalanchoe pinnata*, brought from Africa in the distant past and which is popular as a pot plant, avoid water stress by closing off the tiny breathing pores in their leaves during the day.

Forests on Ultrabasics

On the east and south-east peninsulas of Sulawesi lie the world's largest tracts of ultrabasic rocks, rich in iron, magnesium, aluminium and heavy metals. The soils that develop on these rocks are uncommonly infertile, something which is known by the indigenous peoples and the reason why much of the area still has forest growing on it. It was less well known among central planners until some development schemes targeted for these 'empty' lands failed spectacularly.

Compared with other lowland forests, Sulawesi ultrabasic forests have rather short trees and appear to be dominated by the myrtle family (to which gum and clove trees belong). Epiphytic plants such as *Hydnophytum* and *Myrmecodia* which have relationships of mutual advantage with ants are quite common, as indeed they seem to be in other relatively unproductive forest types. This may be because the nutrients available to the epiphytes in these forests are inadequate in quality and/or quantity, and additional means of obtaining 'food' need to be found.

With the exception of ants, animal life in ultrabasic forests, such as parts of Morowali National Park, is in general very limited. This may be due to the low production of fruit and leaves, or to high levels of toxins in the leaves.

Monsoon Forest

The eastern side of Palu Bay in Central Sulawesi is the driest place in Indonesia, with a rainfall of only 500–600 millimetres (20–25 inches) each year, and its natural vegetation is monsoon forest. Rather more than 250 years ago the valley was described as a 'blessed place', but sixty years ago a colonial officer opined that 'I also think that nowhere in the Archipelago has deforestation had such a fatal influence as in this place ... Centuries of using fire to encourage the type of forage favoured by sheep, goats and cattle has now reduced much of the area to a dull grassland'. Fire is a very potent agent in dry areas such as this. There is one patch of much-disturbed monsoon forest left in the small Paboya Reserve, a short drive from Palu, but the relentless collection of sandalwood from it leaves little of interest. Elsewhere, dry grassland predominates. Perhaps the most interesting plant here is the large Prickly Pear Cactus (*Opuntia nigricans*) brought from South America at the end of the last century. By 1911 this objectionable plant was growing in dense profusion in abandoned ricefields, and twenty years later it was the dominant plant in the region. In 1934 the Cochineal Mealy Bug (*Dactylopius tomentosus*) was introduced to Palu from Australia where it had scored a certain degree of success in controlling the same Prickly Pear there. Just five years later the bug had done its work in Palu and the plant remained only in small pockets. Today, it is very common again over quite a large area, particularly around the campus of the new university, and it seems that a fresh introduction of the mealy bug is required.

The Palu Valley in Central Sulawesi is the driest place in Indonesia with an annual rainfall as low as 300 millimetres (12 inches). The vegetation is very different from most of the rest of Indonesia, and one of the most characteristic plants is the introduced Prickly Pear Cactus (*Opuntia nigricans*). A scale insect successfully controlled its spread in the 1930s and 1940s, and it appears that a new release of the insect is now required.

Mammals

One of the best reasons for a biologist to go to Sulawesi is to see some of its remarkable and very distinctive mammals. Almost all of the non-bat species are found only on Sulawesi, and a good number of the bats, particularly the fruit bats, are similarly restricted. One of the most attractively marked Sulawesi mammals is Wallace's Fruit Bat (*Styloctenium wallacei*) which has a beautiful brown-and-white striped face and back. A good place to see this is in certain North Sulawesi markets where bats are sold for food. The better-known mammal species are the Babirusa, macaques, Cuscus, Tarsier and Anoa, and these are described below.

Babirusa

It's a pig, surely? The enigmatic Babirusa (*Babyrousa babyrussa*) has had no common ancestor with pigs for 30 million years and in many ways is as similar to a hippo as to a pig. To confuse matters further, its name means 'pig-deer' in Indonesian, referring to the curly tusks of the males, but its similarity to a deer does not really stand close examination.

Next time you view one of the grotesque statues of the Balinese demonic man-beast Raksasa, look carefully at its teeth. In most cases you will see that it has curved tusks piercing each cheek: a characteristic of the Babirusa and no other living animal. It seems likely that early Buginese traders from Sulawesi may have brought, or at least brought knowledge of, Babirusa to Bali and the peculiar head was then incorporated into the design of the statues and images.

There are a number of ways in which Babirusa are ecologically dissimilar from pigs. For example, they do not dig around in the soil for roots and worms but rather eat fallen fruit and break open fallen tree trunks to obtain beetle larvae. They also differ in their reproductive strategy. Pigs have large litters of fast-

The Babirusa (*Babyrousa babyrussa*) is a peculiar pig-like animal endemic to Sulawesi. Early writers believed that the male Babirusa used its curved tusks to hang from a branch when it was tired, but it is now realized that these are used in combat with other males.

growing young, whereas Babirusa produce only one or two offspring at a time and these grow relatively slowly.

The peculiar-shaped tusks of the male Babirusa often provoke discussion as to their function. In the past, naturalists believed that the animal used them as hooks to rest its head on a branch or, less fancifully, that the tusks served to protect the eyes as the Babirusa foraged for food among spiny plants (though the female's need for protection seems to have been ignored). Alfred Russel Wallace wrote in 1869 that he was 'inclined to believe rather that these tusks were once useful, and were then worn down as fast as they grew; but that changed conditions of life have rendered them unnecessary, and they now develop into a monstrous form'.

It seems now that the use of the horns in male-to-male combat was not considered. Close examination reveals that the lower tusks are deliberately sharpened against small trees, although this must take a considerable time. The upper tusks are often chipped, scratched or broken; thus it appears that neither tusk has lost any of its usefulness. When males spar they push against each other with their shoulders, rearing up on their hind legs and jabbing upwards with their potentially lethal lower tusks to gore their opponent. The skill is for one boar to hook its upper tusk over the opponent's lower tusk, thereby disarming it but still allowing the one with the advantage to jab or cut the other's neck, face or throat. The spiral shape of the upper tusk is such that, once caught, it is very difficult for the other animal to escape.

The Babirusa is rare and is becoming rarer despite full legal protection. Dozens of carcasses are brought to the markets in North Sulawesi every month, albeit sometimes disguised as wild pig meat, causing the species' already restricted range and density to be further reduced.

Sulawesi Macaques

Not all species of Sulawesi plants and animals are found throughout the island, even if apparently suitable habitat is available. This reflects geological history, climate and perhaps

BELOW There are four geographically separated species of black macaques on Sulawesi. This is the Tonkean Macaque (*Macaca tonkeana*) which is found from the mountains around Enrekang in South Sulawesi, north to Gorontalo in North Sulawesi.

the influence of man. Among the animals, the most striking example of this is found among the monkeys. There are four species on Sulawesi, three with two distinct subspecies, all looking quite similar to the brown and grey Pig-tailed Macaque (*Macaca nemestrina*) of western Indonesia, but differing in being dark brown or black, having a head crest (in the northern species), and in the patterns of greys and browns on their hind-legs. The species in the southern part of the south-west (*Macaca maura*) may once have been cut off from the rest of the island when the sea level was higher and there was a channel between Parepare and Sengkang through the shallow Lake Tempe. Although that was many thousands of years ago, people in that area still tell stories of being able to sail across the peninsula. During that time the monkey populations would have evolved slightly differently, with the result that when the land was connected again, it was no longer possible for them to interbreed with the main population. It is possible to work out similar scenarios for the junctions between the other types of macaques.

Cuscus – Tree Bears

Sulawesi is home to two very different species of cuscus – chunky, tree-dwelling marsupials. The dark-brown Bear Cuscus (*Ailurops ursinus*) is about a metre (3 feet) long with tail and is active during the day when it moves slowly through the tops of

ABOVE The Dwarf Cuscus (*Strigocuscus celebensis*) is the smaller of the two cuscus species on Sulawesi.

the trees eating leaves. It is probably the most primitive of all living cuscus and possums. The Dwarf Cuscus (*Strigocuscus celebensis*), as its name implies, is only half the size of its relative and has very different habits. It is nocturnal, eats primarily fruit and insects, and is much more often seen in the lower canopy. It could not be described as swift in its movements, but can move fairly rapidly when required.

Tarsier

Probably the most endearing member of the Sulawesi forest community is a tiny, goggle-eyed primate called the Spectral Tarsier (*Tarsius spectrum*). With a head and body just 10 centimetres (4 inches) long, a wispy tail twice as long again, and

The nocturnal Spectral Tarsier (*Tarsier spectrum*) is one of the smallest primates in South-east Asia. It lives in small family groups and the adult pair sing a duet just before dawn which reinforces their pair bond; this is essential if the young are to be raised successfully.

weighing only 100 grams (35 ounces), it fortunately inhabits the lower levels of the forest and so can sometimes be seen. In fact, tarsiers can be found in many types of habitats including towns, secondary forest, mangrove forests and montane forests. They can be seen at night if a torch is held at eye level and the observer watches for pairs of red reflections. Occasionally a tarsier can be seen springing from trunk to trunk, but most sightings are made as tarsiers rest or stalk their prey of insects, lizards or anything else of that size that might be edible. Tarsiers are most active in the hour or so before dawn when the adult pair and their offspring head back to their sleeping sites in thickets, among vine or fig tangles, or in tree holes.

Just before retiring the entire family sings a complex call notifying all those with an interest in the matter that they are still alive and defending their territory, the boundaries of which they will have marked out with drops of odoriferous urine during the night. There is apparently a wide array of tarsier languages and dialects throughout Sulawesi. The tarsiers at Tangkoko Reserve begin their call with the male giving a regular series of squeaks, which the female joins with a descending series of squeals which then rise in pitch and pace to reach a climax. The similarities between tarsiers and gibbons are marked; they are both monogamous, territorial, give duetting calls, have specialized forms of locomotion, eat nutrient-rich and reliable foods, and have relatively long life-spans (about ten and twenty-five years respectively). These traits allow a single young to be born and looked after with great care and a high probability of survival, and minimize conflicts with neighbours because the calls provide groups with the necessary information to assess one another's fitness. In addition, time is not wasted either looking for or impressing potential mates, or searching for food within a large and poorly known home range. The down side of the arrangement is that excellent sources of food outside the territory cannot be exploited without the risk of conflict with the 'owner', and the loss of any member of the family, particularly an adult, is keenly felt and can totally disrupt the group.

Anoa

The Anoa is a diminutive, dark buffalo endemic to Sulawesi. Some authorities hold that there are two species: the Lowland Anoa (*Bubalus depressicornis*) which is relatively small, standing about 75 centimetres (30 inches) at the shoulder, with a short tail, and smooth round horns, and the Mountain Anoa (*Bubalus quarlesi*) about 1 metre (3 feet) at the shoulder with whitish legs and rough horns which are triangular in cross-section. Unfortunately the distinction is not rigid and it is not unknown to find Lowland Anoa in the hills and vice versa. It has been suggested that the difference in horn shape is simply a function of age.

Local people with good knowledge of the forest agree that the unpredictable behaviour of the Anoa and its sharp short horns makes it the most feared animal in the region. It is periodically suggested that the Anoa would make a good domestic animal, especially if crossed with the domestic Water Buffalo (*Bubalus bubalis*). Both zoos, and the Toraja people who have kept Anoa for meat, will attest to the unfortunate fact that these animals just do not appear to tame down, however long they are kept.

Anoa eat young leaves and fruit, grasses and ferns. This type of diet causes problems for a herbivore because it contains very little of the essential element, sodium. Anoa have been seen to drink sea-water, but those inland must find some other source of salt, and hence frequent mineral licks. Anoa are exceptional among the five species of wild cattle in South-east Asia in that their major habitat is undisturbed forest, rather than forest edges and secondary growth. This requirement for mature forest makes them unable to adapt to logged areas.

Although there are no figures for the number of Anoa remaining in Sulawesi, it is generally agreed that they must be declining due to the combined effects of hunting and forest loss. They are protected by law and many of the areas they live in are within reserves and national parks, but the regulations enacted to ensure their survival are flouted and the judiciary have yet to take offences particularly seriously.

Mountain Rats

Sulawesi mountains are also home to a number of rats and shrew-rats, so called because of their relatively long snouts. These forest rats are quite distinct from the pestilential and despised rats of towns and agricultural areas, for in their stable environment they give birth to just one or two young at a time. Some of these species appear to have very restricted ranges, occurring only in a narrow altitudinal band on the mountain tops. These live among the mossy maze of roots and fallen trees in the upper montane forest. As with other communities of rather similar animals, there is a clear segregation between the species in time, space and diet: some are active during the day, some at night; some live entirely on the ground whereas others are partly arboreal; and some eat only fruit, some only insects, some only worms, some leaves, and some a variety of foods.

Market Snacks

Markets in eastern North Sulawesi are interesting for some of the unlikely meat on sale. The stalls of wild pig meat are perhaps unremarkable, but the oft-seen hunks of Babirusa meat are disturbing because this threatened beast is fully protected by Indonesian law. The yelps of small dogs are best not investigated because the animals are probably destined for a dinner plate. Also on sale are what look at first glance like sausages on sticks. A tail dispels this illusion. A number of species of rats and even the Dwarf Cuscus are brought to market barbecued on a stick, and make popular snacks. A good number of the larger species of fruit bats are sold too, though generally uncooked.

ABOVE Market scenes are colourful and interesting throughout the archipelago, but in the Minahasa region of North Sulawesi they are characterized by the presence of barbecued forest rats, and large fruit bats ready for cooking.

Endemic Birds

Of Sulawesi's 247 known resident birds, an amazing 36 per cent are endemic including seventeen endemic genera. New records of species are still being made and it is quite possible that new species await discovery, particularly in the mountains of South-east Sulawesi which have received very little attention. As on the other islands, large and conspicuous birds are rarely seen above about 1,200 metres (3,950 feet), the most common exceptions being eagles and other birds of prey. Above this but below the highest peaks, the bird fauna is characterized by a greater relative abundance of pigeons and doves which include the very rare Dusky Pigeon (*Cryptophaps poecilorrhoa*), and of songbirds such as the Serin (*Serinus estherae*). Among the more unlikely birds in this area is the mottled-brown and secretive Sulawesi Woodcock (*Scolopax celebensis*), which is known from fewer than ten museum specimens. The endemic species include the Crowned Myna (*Basilornis celebensis*), the long-tailed White-necked Myna (*Streptocitta albicollis*), the iridescent, dark-green Purple-bearded Bee-eater (*Meropogon forsteni*), the brightly coloured Red-knobbed Hornbill (*Rhyticeros cassidix*), the shy and dapper black-and-white Piping Crow (*Corvus typicus*), the Finch-billed Starling (*Scissirostrum dubium*) which is most striking for the huge colonies which nest in holes bored out of large dead trees, and the famous Maleo (*Macrocephalon maleo*).

Maleo

The Maleo is the largest and most striking member of a small family of chicken-like incubator birds or mound builders found in eastern Indonesia, Australia and Polynesia. It may once have been found throughout the island in suitable habitats, but man's predations and habitat loss have restricted it now to areas of North, Central and South-east Sulawesi. Maleos are about the size of a domestic hen with a black erect tail and back, rose-tinted white belly and a helmeted head. They are primarily birds of the lowland forest but at breeding times they search for warm ground in which to lay their eggs. These communal

ABOVE A colony of Finch-billed Starlings (*Scissirostrum dubium*) nesting in a dead tree near Tangkoko Reserve in North Sulawesi.

breeding grounds may be hot beach sand or the soil around volcanically warmed vents or hot springs.

Sounding like a cross between a duck and a turkey, an adult pair will busy themselves on the ground looking for somewhere to dig their nest, testing the suitability and temperature of the soil. When both birds are satisfied, they join together to dig a deep pit using their large feet and claws. Their toes are slightly webbed which makes their scratching more efficient. After about three hours of intermittent digging and resting, the female lays an egg. The pit is then refilled until the female is ready to lay her next egg about ten days later. The process is complicated by a number of decoy pits dug by the birds to distract predators such as monitor lizards away from the real nest.

Man has sought out these eggs as delicacies since time immemorial, for they are no ordinary eggs. The Maleo lays proportionately one of the largest eggs of any bird. Each is 11 centimetres (4 inches) long and weighs more than 250 grams (9 ounces), more than twice the length of a chicken's egg and about five times the weight. In historic times the trade in Maleo eggs was controlled by piratical rajas within whose petty states the nesting grounds lay. Wallace watched hundreds of Maleos at a communal nesting ground on a beach next to the present-day Tangkoko Reserve, from which eggs were taken in a sustainable manner. However, within six years of a settlement being established behind the beach in 1911, the nesting site had been abandoned. At another site on the south of the northern peninsula the number of eggs laid now is less than 10 per cent of that fifty years ago. Although trade is illegal today because the Maleo is fully protected, it is unfortunately possible still to find individual eggs wrapped in palm-leaf baskets for sale in Jakarta markets. This is another threat to a species which is already under severe pressure because of loss of lowland forest and of the specialized nesting grounds.

It has been demonstrated that Maleo eggs can be collected and reburied in secure, fenced enclosures and still hatch out an acceptable proportion about three months later. The young hatchlings already have adult plumage and 'explode' out of the ground, flying towards the nearest forest cover, after perhaps three days of burrowing upwards. Various management techniques are feasible, but they all depend on the security of large areas of relatively undisturbed lowland forest in which the birds can live.

Fishes

It is only recently that the uniqueness of Sulawesi's freshwater fishes has been brought to the attention of people outside the corridors of ichthyological endeavour. An entire family, the duck-billed fishes or Adrianichthyidae, is restricted to the two lakes of Lindu and Poso in Central Sulawesi. Lake Poso has three of the four species. The female of one of these remarkable species, *Xenopoecilus oophorus*, carries its eggs attached by thin threads to its belly until they hatch.

The Malili Lakes at the north-east corner of South Sulawesi are even more remarkable; within this complex of lakes, rapids and rivers have evolved no fewer than sixty species of crabs, shrimps, snails and fishes, many of which are extremely colourful. Strangely, only one species is common to all four lakes, and that is a shrimp. The second largest and most accessible lake, Matano, is remarkable physically, too, because it is extremely deep, 540 metres (1,770 feet) but its surface is only at 380 metres (1,245 feet) above sea level, so that the bottom of the lake is actually 160 metres (525 feet) below sea level. Both the nickel mine near the shores of Lake Matano, the increasing human population pressure and the desire to develop a commercial fishery on the lake pose serious threats. The environmental management of the mine area is conducted with considerable care, and the physical properties of the lake probably will preclude a successful commercial fishery, but attempts to establish one could bring in persistent parasites and diseases.

Invertebrates

The knowledge of Sulawesi's invertebrates has increased dramatically over the last few years, thanks mainly to Project Wallace organized by the Royal Entomological Society of London. The royalty among the thousands of species are the butterflies, the kings of which are the Swallowtails.

Swallowtails

Swallowtails are among the most dramatic of butterflies and thirty-eight species are known from Sulawesi, eleven of them endemic. Some are extremely rare, known from just a few specimens in museums. One species has been the cause of many visits by lepidopterists to Sulawesi because it was made famous by Wallace in his *Malay Archipelago* book. It concerns a Swallowtail known as *Graphium androcles* which he saw at the Bantimurung waterfall near Maros, South Sulawesi. He wrote, 'As this beautiful creature flies, the long white tails flicker like streamers, and when settled on the beach it carries them raised upwards, as if to preserve them from injury. It is scarce, even here, as I did not see more than a dozen specimens in all and had to follow many of them up and down the river's bank repeatedly before I succeeded in their capture'. That and other species of butterflies in classic haunts are extremely scarce now because of general collecting for the tourist curio trade, and also because of the specialist trade whereby certain species command very high prices on the hidden market.

Coconut Crabs

On the remotest islands in Sulawesi and east through the Moluccas to the Pacific can be found the largest of the world's terrestrial arthropods. This is the Coconut Crab (*Birgus latro*), a relative of the hermit crabs. These enormous crabs weigh up to 5 kilograms (11 pounds) and their outstretched claws can span 90 centimetres (35 inches). Unlike most crabs, the Coconut Crab mates on dry land, and the female carries the fertilized eggs down to the sea when the tides are highest, around the full moon. The newly hatched larvae are washed away and for one to two months float in the plankton before finding a small shell to live in and migrating to the nearest land. When they are about two years old, about 2.5 centimetres (an inch) across and like miniature adults, they give up trying to find shells. The crabs then hide in holes or under vegetation during the day to avoid being dried up by the sun. At night they come out and forage for cracked coconuts and pandan fruits, their major foods, but are catholic in their tastes and will eat snails, unwary birds, leaves and even other crabs.

The coconut and the Coconut Crab seem to have an extremely close relationship; the mature crab larvae may ride on floating coconuts in the sea until they are cast by the waves onto a distant beach. Thus, before man took a culinary interest in both the crab and the coconut, the distributions of the two were probably very similar. That pattern has now become blurred with the one being overexploited and the other cultivated.

On some islands the Coconut Crab is the largest creature present, but it fares badly in close proximity with man because it is delicious and easy to catch and sell, sometimes in enormous quantities. In the Togian Islands, for example, it is eaten locally, offered to visitors, and exported both to the Moluccas and to other towns in Sulawesi. Not surprisingly, this is an internationally threatened species, although its legal protection in Indonesia seems unable to stem the tide of exploitation.

The early travellers' tales concerning these extraordinary animals were somewhat fanciful and not, it seems, based on direct observation. No one has yet demonstrated the crabs' ability to climb coconut trees, pinch through the fruit stem, and scuttle back down the trunk to open and savour the fallen fruit.

Coconut Crabs (*Birgus latro*) are the world's largest terrestrial arthropods. They are becoming increasingly rare in Indonesia as they are caught to grace the tables of discerning but unthinking diners in the eastern half of the archipelago. These crabs are nocturnal but are easily caught using torchlight.

Extinctions

Many endemic Sulawesi animals are believed to be at risk and it is possible that a few species have even recently dropped into the abyss of extinction. Among the well-known threatened species are most of the larger endemic mammals such as the Spectral Tarsier, Sulawesi Civet (*Macrogalidia musschenbroecki*), Babirusa and Anoa, as well as the Maleo, and butterflies such as the Palu Swallowtail (*Atrophaneura palu*), the Talaud Black Birdwing (*Troides dohertyi*) and Tambusisi Wood Nymph (*Idea tambusisiana*). In addition to these are entire communities of fishes and other animals in the Malili lakes and other Sulawesi lakes.

One of the bird species which is probably now extinct is the Caerulean Paradise Flycatcher (*Eutrichomyias rowleyi*) which is known from but a single specimen caught in 1873 on the island of Sangihe between North Sulawesi and the Philippines. Since then virtually all of the Sangihe forest has been converted into coconut and nutmeg plantations. Two expert ornithologists have tried to locate this bird in and around the small patch of forest remaining on the top of the highest mountain, with only one of them making a possible unconfirmed sighting of a single bird. There are three other endemic bird species on Sangihe, two parrots and one sunbird, but these seem able to adapt to secondary vegetation and have been sighted recently.

Another species which may be extinct is the goby fish *Weberogobius amadi* from Lake Poso in Central Sulawesi. Recent efforts to collect the fish have been unsuccessful. Nearly every specimen of the four endemic species of fishes in the lake caught in 1989 had parasites clinging to them, fungal growths, or wounds, and those viewed underwater were mostly swimming with difficulty or clearly dying. It is likely that diseases and parasites were inadvertently introduced along with a popular catfish and predatory snakehead fish when government-sponsored migrants from Java were settled in the area. One can only hope that the same fate does not await the ten or more beautiful endemic species of fishes in Lake Matano where a number of food fishes have already been introduced.

Regrettably, such extinctions are likely to continue being recorded as we gather more information on the distribution of animals, and as the basic problems facing natural ecosystems remain unsolved. It is important, of course, not to cry wolf over the lack of records of certain species because no one may have gone out looking for a particular endemic animal or plant for many decades. That is certainly the case for the endemic birds and mammals of the mountains in South-east Sulawesi which have not been seen by biologists for over forty-five years. But where biologists have looked, and in places where the natural habitats have all but disappeared, genuine extinctions are likely to be detected.

The Moluccas

Widely scattered tropical islands with clean sandy beaches and extensive coral reefs make up the Moluccas. Ambon, so picturesquely described by Alfred Russel Wallace in the mid-1850s, has now become virtually an urban sprawl; but much of the province has a very low population density and some of the smaller islands are uninhabited save for birds and occasional visiting fishermen. Here the sea is a dominant force and influence in the lives of many of the inhabitants, although in the interior of the large islands there are groups of people who live quite independently of the sea. There are many wild areas in the province and, if the problems of transport can be overcome, then there is much to discover and enjoy. The forests and their animals were little known in Wallace's day and, in many places, particularly on the large island of Halmahera, this is still the case.

Forests and Trees

The vegetation types of the Moluccas are highly influenced by climate and topography. In the mountains of Seram, Obi, Morotai, Buru and Halmahera there are verdant montane forests and a good range of natural habitat types, and it is on Seram that the single most important conservation area in the province can be found, Manusela National Park with its wide range of forest types. Only in north-west Halmahera and on Seram are there everwet lowland forests, and these are being exploited by logging companies, primarily for the valuable damar trees (*Agathis*), close relatives of the famous New Zealand *kauri* trees. Their first-class timber is used for cabinet making and interior strip panelling. They used to be tapped widely for their resin which was used in the manufacture of spirit varnishes and lacquers, but this has decreased somewhat with the availability of resin from pine plantations.

In the drier areas the Paperbark Tree (*Melaleuca cajuputi*) grows in nearly pure stands. This species is able to survive the regular fires of the dry seasons by virtue of its loose-fitting, papery bark. It is extremely common on the former prison-island of Buru, and supports an industry of cajuput oil production. The oil, which is distilled after boiling up pounded young leaves, is valued for its medicinal properties and is exported to be used in medication for the relief of muscular pain and headaches.

One of the naturally occurring pioneer trees in Moluccan disturbed forests is the Batai (*Albizzia falcata*) whose range encompasses the Philippines and the Solomon Islands but which has been planted widely elsewhere in South-east Asia. It has proved itself a wonderful plantation tree and some

There are hundreds of small coral-fringed islands in the Moluccas, many of them uninhabited. She-oaks (*Casuarina equisetifolia*) border the beach in this remote island in the Kai group, south-east of Seram.

LEFT View to Hiri Island from Ternate, the seat of a once-powerful Sultanate that ruled from eastern Sulawesi to the Moluccas.

BELOW Dugongs (*Dugong dugon*) feed mainly on the carbohydrate-rich rhizomes of seagrasses, found just below the surface of the sand or sediment, which they 'chew' using the horny pads at the front of their jaws. They do not have to compete with other animals for this food, but their existence is threatened by man, who has been hunting them with increasing intensity ever since his boats and weapons made it possible.

seedlings not quite two years old have reached over 16 metres (55 feet); the fastest tree growth known. Its wood is used to make veneers, but the species is used most frequently for planting up bare ground and as a shade tree for other trees which are less tolerant of full sun. Another similarly useful tree indigenous to the Moluccas (and Sulawesi) is *Anthocephalus macrophyllus*.

Eastern Indonesia is also home to the only species of *Eucalyptus* to be found in rain forests, *Eucalyptus deglupta*, which has peeling bark giving the trunk an array of rich pastel shades. This tree can reach an awe-inspiring 78 metres (255 feet) tall and it seems incredible that these giants start as infinitesimal seeds. They tend to grow on bare ground such as landslips and old river banks.

Mammals

The native land mammals are restricted to a few species of rats and bats, some of which are endemic, on the larger islands. The other mammals such as cuscus, monkeys, civet cats, pigs and deer were almost certainly introduced by man for food hundreds of years ago, and cats and rats have reached many of the smallest islands. The Babirusa is found on Buru, but this was probably introduced from Sulawesi. The largest native mammal in the Moluccas is the Dugong (*Dugong dugon*) which lives in shallow coastal waters.

Dugong

The Dugong is the only herbivorous marine mammal and used to be found in coastal waters from east Africa to the south Pacific. Although it is superficially similar to a seal, it is more closely related to the elephants, which are also thought to have been swimming, shoreline animals at an earlier stage of their evolution. Just how true it is that the mermaids of fable are based on the crazed imaginings of sailors who have set eyes on Dugongs is open to debate. Dugongs are only a little longer than humans, and suckling females do have large mammary glands

in about the right place; but their heavily whiskered faces, their failure to bask seductively on rocks, and their inability to hold mirrors in which to view their bald heads argue against them being the origin of the stories.

Like elephants, Dugongs have relatively inefficient digestive systems and must eat very large quantities of plant material to extract enough nutrients. They feed in the seagrass meadows which are often found on shallow sandy plains near coral reefs. They eat seagrass leaves but the bulk of their food is the carbohydrate-rich rhizomes, found just below the surface of the sand or sediment, which they 'chew' using the horny pads at the front of their jaws. Their lives should be relaxed and easy since no other creatures feed on the same food – the closest competition coming from the Green Turtle (*Chelonia mydas*) which eats just the leaf blades of the seagrass. The ease of their lives is thwarted by man, however, who has been hunting them with increasing intensity ever since his boats and weapons made it possible. Dugongs provide not just meat but also oil and tusks. Unfortunately they are easy to catch and ill-equipped to deal with this hunting pressure because they reproduce slowly. Dugongs are at least ten years old before they are mature and they give birth to a single young after a one-year gestation; there is then usually three to seven years before the next pregnancy. Apart from falling prey to deliberate hunting, Dugongs are also caught accidentally by fishermen who are quick to capitalize on their gain, despite the legal protection of this species.

Birds

The Moluccan islands have about 350 species of resident birds of which eighty-nine are endemic. Halmahera, Bacan and Morotai have the most bird species, with 126, and also the highest level of endemism, 21 per cent, including six species of parrots. The largest bird in the province is the Two-wattled Cassowary (*Casuarius casuarius*), a bird more generally associated with New Guinea, which lives in Manusela National Park, while the most beautiful is arguably the male King Bird of Paradise (*Cicinnurus regius*) from the Aru Islands which has a whiter-than-white belly, blue legs, a ruby-red head and back, and two thin wires leading from the edges of the tail which end in iridescent green discs.

Parrots

The Moluccan islands are justly famous for their parrots which total no fewer than thirty-one species with the most striking plumage and colour combinations. Of these, perhaps the most popular are the cockatoos, such as the delicate pink Salmon-crested Cockatoo (*Cacatua moluccensis*) known only from Seram and a few small neighbouring islands. The cockatoos are too attractive for their own good as they suffer grievously from trapping to supply the apparently insatiable demand from within and outside Indonesia for pets, resulting in tens of thousands of parrots being exported from the Moluccas each year, many of them dying *en route*.

Most parrots are birds of lowland forests, moving around in small groups or large flocks seeking fruiting or flowering trees. The smaller species such as lories and lorikeets eat mainly soft fruit, nectar and pollen, using their specially adapted tongues. The larger species such as the cockatoos are strong enough to open up the fruits of palms and kenari to reach the softer seed inside, which humans find singularly inaccessible without the aid of saws or heavy stones. The Salmon-crested Cockatoo is also able to open young coconuts and is regarded as a pest in those areas where bird traders have not yet seriously reduced the populations. Parrots are very noisy and colourful and thus very conspicuous when they raid crops; they therefore receive a disproportionate amount of blame as perceived pests – while the silent and nocturnal rats and squirrels probably do at least as much damage.

The Eclectus Parrot (*Eclectus roratus*), a widespread species known from the Moluccas through to islands east of New Guinea, is one of the few parrots in which the males and females can be easily distinguished: the male is green while the female is blue and red. For nearly a century the two sexes were thought to be two completely different species. Other surprising species are the green-coloured Black-lored Parrot (*Tanygnathus gramineus*) which has a striking black line from its forehead to its eyes, and is restricted to Buru Island, and the small species of Hanging Parrots, such as *Loriculus amabilis* of Halmahera and smaller islands to the south-west, which hang upside down like little green bats. They have very short, rounded tails, and the females have the unusual habit of carrying nesting material

The Rainbow Lory (*Trichoglossus haematodus*) has a wide distribution from Bali through the Moluccas to New Guinea, and northern, eastern and south-east Australia. It feeds on pollen, nectar, fruit, seeds and insects. These birds are also said to raid fields of maize to feed on the young seeds.

among their rump feathers. These attractive little birds do not fare well in captivity, although they often live long enough to be sold in bird markets.

Bird Islands

Some of the smallest, uninhabitable pinpricks of islands are of considerable significance as breeding sites for large numbers of seabirds such as frigatebirds, tropic birds, boobies, terns and other smaller species. Diving groups of these birds have for centuries attracted Indonesian fishermen to shoals of fish near the surface of the sea, but all the indications are that they are becoming less abundant as predatory rats and cats are accidentally released onto the islands, and as fishermen remove eggs from nests to sell. On Suanggi Island near to Banda, for example, the Greater Frigatebird (*Fregata minor*) no longer breeds and the Red-footed Booby (*Sula sula*) and Brown Booby (*Sula leucogaster*) nest only in the remote corners.

The most amazing of the bird islands are the two nature reserves of Manuk Island and Mount Api in the Banda Sea; they are the breeding and roosting sites for millions of seabirds and are probably the greatest of the bird islands left in the whole of South-east Asia. Both these islands are dormant volcanoes but Manuk has much more vegetation because it is longer since it last erupted. Manuk is also more hospitable and fertile than Mount Api, with the result that fishermen-turned-farmers have illegally cut gardens on the slopes, thereby in places ousting both the tree-nesting Red-footed Booby and Greater Frigatebird, as well as the ground-nesting Masked Booby (*Sula dactylatra*).

The Greater Frigatebird is an argumentative creature and will mob other species while they are hunting for fish in order to make them regurgitate or drop their catch, which the frigatebird then retrieves for itself. Mostly, however, these birds are seen soaring high over the sea or coasts searching for surface-feeding shoals of fish. The shiny black males are a magnificent sight as they inflate their enormous strawberry-red throat sacs to display

Some 300–400 pairs of Red-footed Boobies (*Sula sula*) and a few pairs of Brown Boobies (*Sula leucogaster*) breed on Suanggi Island, and a few hundred pairs of Greater Frigatebirds (*Fregata minor*) roost there. Suanggi, one of the Banda group south of Seram, is a tiny volcanic island 500–1,000 metres (1,500–3,000 feet) in diameter. It is a very important seabird island and is surrounded by extensive coral reefs. Despite its small size, isolation and lack of fresh water, fishermen have removed much of the original vegetation and planted crops, and they have inadvertently introduced rats and cats which are a major threat to the survival of the seabirds.

to females flying overhead. They begin breeding when they are seven to eight years old, and the females appear to breed only every other year. This slow rate of increase makes the frigatebirds very vulnerable to increased pressures of any kind. They nest in the same trees as the Red-footed Boobies on Manuk, but conflict is reduced because the peak breeding times of the two species are different.

One of the most beautiful birds in the region is the exquisite Red-tailed Tropic Bird (*Phaeton rubricauda*) of the tropical Indian and Pacific Oceans. It is largely white with black tips to its wings and a black strip from its eye to its beak, but has a striking red bill, and tail streamers which double the overall length of the bird. Somewhat fewer than a thousand tropic birds breed on both Manuk and Mount Api, but on Manuk they are forced to seek out nesting sites which the introduced rats cannot reach. There are not enough of these safe sites and the population of tropic birds has no chance of recovery unless the rats are exterminated on the islands.

Reptiles

The Moluccas has a relatively rich reptile fauna and these animals play an important role in forest ecology in the absence of many mammalian forest inhabitants. For example, the monitor lizards and pythons are the dominant large predators in the forests, and a variety of lizards probably take over some of the ecological roles filled elsewhere by squirrels, tree-shrews and civet cats. The largest reptiles, such as crocodiles, large pythons and monitor lizards, are sought for their saleable skins, and the water-walking Sail-fin Lizards (*Hydrosaurus amboinensis*) are caught to be stuffed and sold as curios (despite being legally protected). Halmahera, Bacan and Morotai have between them nine endemic species of reptiles. One of the snakes from Bacan

in fact represents an endemic genus, *Calamorhabdium kükenthali*, distinguished by its white collar and a very short tail which is just one-twentieth of its total length of 20 centimetres (8 inches). Another endemic species of snake is *Xenochrophis halmahericus* which is locally called a 'flying snake' and is said to be able to glide between trees. In addition, the remote sandy beaches of the Moluccas are important nesting sites for five species of marine turtle.

Turtles

The largest species of turtle found in Indonesia is the enormous Leatherback (*Dermochelys coriacea*) which has a dark-brown, ridged carapace up to 2.5 metres (8 feet) long and can weigh up to a ton. The other four species are all about the same size: 90–100 centimetres (35–40 inches) long and weighing 80–100 kilograms (175–220 pounds). Of these, the Olive Ridley (*Lepidochelys olivacea*) in the west and Loggerhead (*Caretta caretta*) in the north occur in very small numbers with no nesting sites known for the latter, and a few tens of thousands of Hawksbills (*Eretmochelys imbricata*) are found throughout the archipelago. It is not known to what extent the difference in abundance is entirely natural or man-induced. In contrast, the relatively common Green Turtle ranges and nests throughout Indonesia. It is, however, under continual pressure from egg collecting and the demand for its meat and shell. In addition, the Green Turtle has to face natural threats such as storm surges washing away its eggs, and wild pigs and monitor lizards eating its eggs and nestlings. Green Turtles are protected in almost every country in the world except Indonesia, reflecting the very large numbers present within its boundaries. Even so, the large numbers of adults taken from the Moluccas, the easy availability of eggs in most major markets, the knowledge of many nesting beaches which are no longer, or much less frequently, visited by turtles, and the absence of unequivocal information on the effectiveness of current management programmes, all argue for the greatest caution lest these defenceless animals be pushed to the brink of extinction or beyond.

Turtles all look superficially similar yet they feed on a wide range of foods. The Leatherbacks live on a diet of jellyfish and this sustains them on their lengthy migrations across the oceans, during which they swim thousands of kilometres.

Although more often seen hauling themselves up beaches to lay eggs, turtles such as this Green Turtle (*Chelonia mydas*) spend most of their lives swimming in the tropical seas. Green Turtles graze on seagrass and algae growing in sandy and reef areas.

Loggerhead turtles eat crustaceans and molluscs, Olive Ridleys eat crabs and shrimps in shallow seas, Hawksbill turtles eat a range of invertebrates associated with coral reefs, and the Green Turtle is a vegetarian which feeds on seagrass.

A number of management schemes have been in place in Indonesia for some time to try to give Green Turtle eggs and nestlings the best possible chance of survival. Adult females are allowed to haul themselves up the beach and lay the eggs, which are then taken to a fenced hatchery where they are buried in the sand to incubate. After hatching, the young turtles are collected together and taken out to sea thereby avoiding beach predators and inshore fishes. An unexpected complication with this form of management is that the sex of a developing turtle is determined by the temperature during the middle third of incubation, with more females being produced under higher temperatures. Artificial incubation could therefore result in young turtles of only one sex being hatched, with disastrous results in the future. This argues for a number of sites to be used at a hatchery providing different conditions of exposure and proximity to the sea. Not all beaches are suitable for turtle nesting and while the determining factors have yet to be established beyond doubt, it seems that the degree of protection from prevailing winds and the salinity of the moisture within the sand influences the choice of the females.

Sail-fin Lizards

The largest lizard in the Moluccas (found also in Sulawesi, New Guinea and the Philippines) is the Sail-fin Lizard (*Hydrosaurus amboinensis*) which reaches just over a metre (3 feet) in length, only one-third of which is body. It is a shy creature but is most frequently seen basking on a branch overhanging water, with its feet dangling down on either side. The term 'sail-fin' describes the high crest which runs down the length of its tail. This stands relatively erect because it is supported by projections from the backbone.

Sail-fin Lizards are usually found near water and it is intriguing that their toes are equipped with enlarged scales, which are most marked in juveniles. These are used both for swimming, at which they are very adept, and also for running across the surface of the water when danger threatens. The only predators that need be feared by the adult lizards, other than man, are large pythons and eagles; juveniles would make good meals for a wider variety of predators, however, and this may be one reason why they are less frequently seen. They are very wary of man and it is extremely difficult to approach them.

Sail-fin Lizards have the additional distinction of being the only leaf-eating lizards in Indonesia, although they do also take insects and fruit when available. They do not have teeth, so they perforate leaves with the pointed edges of their mouth and then pull a portion off, somewhat like tearing toilet paper. They do not have a gizzard for grinding the leaves either, but they may eat stones occasionally to help break down the food. Nor do they possess specialized stomachs for processing leaves, as do many leaf eaters, so one is tempted to ask why they bother to eat leaves at all. The answer is probably that this easily available subsistence diet enables them to spend long periods waiting motionless for suitable prey to come within reach.

Butterflies and Bees

Among the Moluccas' twenty-five endemic butterfly species are four species of the famed birdwings, enormous butterflies, often with startlingly beautiful wings which have been and still are sought by collectors. Prices reaching thousands of dollars are not unusual for rare species. One of the Moluccan birdwings, *Ornithoptera croesus*, has a wingspan of nearly 20 centimetres (8 inches) in the female and up to 15 centimetres (6 inches) in the male. The smaller male is more brightly coloured, with the top of the forewings shining golden-orange and dark brown, with green and black below. This species was first described by Alfred Russel Wallace in 1858 and is found in the islands of the northern Moluccas where it lives in lowland forests. There are three varieties which differ somewhat in their colour and are found on different groups of islands. With no reserves yet established here and all the lowland forest parcelled out in blocks of largely unsupervised timber concessions, the long-term future of this butterfly gives cause for concern despite its status as a protected species.

Wallace also came across the world's largest bee, *Chalocodoma pluto*, during his time in the Moluccas. It is a formidable creature, with the females growing to 4 centimetres (1½ inches) in length, and is found on the islands of Halmahera, Bacan and Tidore. This bee is remarkable not just for its size, but because it nests communally in inhabited termite nests in lowland forest trees. In order to lay her eggs somewhere safe, the female bee bores a tunnel into the chosen termite nest which is then lined with a mixture of woodchips and resin scraped from wounds in the dipterocarp tree *Anisoptera thurifera*. This lining precludes contact between the bees and termites, and the termite nest appears to be used simply as a substrate for the tunnel. Interestingly, a specialist who spent ten months studying this bee found only seven nests – it thus seems to be singularly rare. When shown a specimen, local people claimed never to have seen it, but they knew of its existence and it had a local name.

Coral Reefs

The large expanse of sea within the province harbours an enormous wealth of marine resources: fishes, marine mammals, turtles, seagrass meadows, large shells such as tritons (*Charonia*) and giant clams (*Tridacna*), and a host of other invertebrates, many extremely beautiful and keenly sought by collectors. Indeed, very few of the coral reef areas in Indonesia, even in the remote parts of the Moluccas, have not known the pressures of organized parties of collectors arriving in well-equipped schooners to take what is saleable – even from the areas designated as marine reserves. One of the most unpleasant forms of exploitation has been of vast numbers of sharks which are caught, deprived of their dorsal fin, and returned to the sea, dead or alive, to be eaten by others. The fins are dried and sold in eastern Asia to make shark-fin soup. Enormous and unselective drift nets have also been used in the seas here, but the government is now trying to put a stop to this practice. There are probably more coral reefs in the Moluccas than in any other part of Indonesia and much remains very good quality. Pearl fishermen, collectors of fishes and invertebrates for aquarium dealers, and other more ordinary fishermen have all left their mark on the reefs, but the damage is not yet irreversible, given both time and enforced protection.

In 1889 Sidney Hickson, an English scientist who spent some years looking at Indonesian reefs, wrote 'A coral reef cannot be properly described. It must be seen to be thoroughly appreciated'. Coral reefs are a whole new world to most land-based biologists since many of the organisms living in a reef are from groups with which humans generally have no contact, such as the colonial sponges and hydroids, sea wasps and sea cucumbers, sea lilies and sea squirts. These are all members of different biological divisions, as far apart in zoological terms as a garden

ABOVE A crinoid or featherstar, a distant relative of starfishes, sea cucumbers and sea urchins.

BELOW A dendrophylliid stony coral; the polyps extend only at night-time.

worm and a bird. Coral reefs are enormously rich in species and more diverse in their membership than even a tropical rain forest.

The coral animal itself starts life as a small planktonic larva which eventually comes to rest on a substrate and changes into a polyp. This begins dividing to form identical polyp neighbours. Each polyp secretes a wall of limestone around it, but each member of the colony is connected to others by thin strands of tissue passing through pores in its hard wall. Within the polyps' skins are small yellow-brown granules which are primitive plants called dinoflagellates. Amazingly, only a single species of dinoflagellate is known to have this special relationship with corals. These plants photosynthesize, releasing oxygen which is used by the polyp, which in turn produces phosphates and nitrate wastes and carbon dioxide to be utilized by the plants. Coral reefs are formed by the compacted and cemented skeletons of sedentary organisms such as corals and some algae which become smothered by the next layer of coral animals and so support the new 'skin'. These organisms need clear, sunlit and warm water and thus coral reefs are not found in deep water or near the mouths of large rivers or urban centres where the water is full of sediment.

It takes some three to eight years for a coral to mature, and most of the corals in an area tend to breed seasonally and simultaneously a few days after a full moon. Breeding corals release a very large number of edible eggs into the sea, providing a bonanza for fish and other animals; this synchronized breeding serves to satiate the predators and prevents them from feeding exclusively on coral eggs or larvae, thus giving the best chance of survival to the most eggs and larvae. It is analogous to the co-ordinated fruiting of dipterocarp trees (see page 30), and similarly applies to a range of species.

Most human visitors to a reef swim or dive during daylight hours and, glorious though the natural spectacle is, gain only a pale impression of the true beauty of a reef. At night, the formerly sharply defined edges of the dull coral shells become soft and multi-coloured as the nocturnal polyps extend their tentacles to filter food out of the water. By day the red, green and yellow wavelengths of the sunlight are quickly absorbed by the water with the result that much of the deeper reef and its organisms appear bluish, but at night, using a torch, the true colours of the reef can be seen in all their splendour.

Conservation

Conservation faces enormous logistical problems in the Moluccas. Many of the islands are remote and inaccessible, their biological resources are inadequately known, those who act illegally in the collecting of biological resources are very efficient and have a high level of technology available to them, and it is often logistically impossible to supervise the activities of those who engage in potentially damaging, yet legal, activities such as logging. Despite the fact that Ambon is the busiest town in the Moluccas and a major focus for air and sea traffic, the large island of Seram next to it is far from easy to reach, has no airport and has little public transport on a skeletal road network. In some ways this is good for conservation, but most of the government's conservation employees are on Ambon rather than in the field marking or guarding boundaries, encouraging the participation of the local people in conservation efforts, or conducting biological surveys.

Despite these many problems, the Moluccas is still a valuable and extremely beautiful storehouse of natural beauty. It is hoped that increasing national and international concern for the state of these terrestrial, coastal and marine environments will focus attention on the problems and encourage the exploration and funding of solutions.

BELOW Dancing children at dusk on a remote beach of Kai Kecil Island in the south-east of the Moluccas.

The Lesser Sundas

Plants

The Lesser Sundas is made up of the string of islands east of Java, although Bali is generally excluded because it is biologically and biogeographically aligned with Java. Most of the islands are thinly populated, dry and agriculturally relatively unproductive, a real contrast to their western neighbours. But if the vegetation seems arid for much of the year, the islands have a sparse beauty of their own.

Plants

The climate, the small size of the islands, and their distance from the Asian and Australian continental blocks, combine to make the flora of the Lesser Sundas very poor in species compared with the larger, wetter islands to the east and west. Australian plants are evident, with the gum trees *Eucalyptus alba* and *E. urophylla* reaching west to central Flores, species of *Acacia* being found throughout (some having had the helping hand of man), and Sandalwood reaching Timor and Sumba. The mountains of the Lesser Sundas do not develop the moist moss forest seen on the larger islands, and yellow-green tassels of the *Usnea* lichen are essentially the only indication that the cloud zone has been reached. To most visitors, however, the most characteristic trees are probably the two tall palms, Gebang and Lontar, individuals of which tower over many areas of grass and scrub.

Forests

The largest block of forest left in the Lesser Sundas is on Sumba, but small pockets can be found on most of the other islands. The majority of these are monsoon forest which has a relatively uniform canopy, few tree species and few climbing rattan palms or epiphytes such as orchids and ferns. Characteristic trees include leguminous species of *Acacia, Cassia, Dalbergia latifolia* and *Pterocarpus indicus*, as well as *Schleichera oleosa*, a relative of the Rambutan tree, which produces a useful hard wood and seeds which yield oil for cooking and lighting. Two well-known introduced trees, the Tamarind (*Tamarindus indicus*) and Teak (*Tectona grandis*), are well suited to the seasonal climate and can now be found in some forested areas. In the Dompu area of Sumbawa there are almost pure stands of the timber tree *Duabanga moluccana* which need to be managed on a sustainable basis. Unfortunately, due to a lack of understanding and knowledge of the government regulations which would allow for such long-term benefits, these may soon be no more than a memory.

East Timor has very few patches of forest left, but some of these contain the last natural stands of very important trees such as the plantation tree *Eucalyptus urophylla* and the Sandalwood. In the inland areas the steep and relatively infertile mountain soils, combined with the long dry season and low rainfall, influence the types of forest present as well as the behaviour

Much of the Lesser Sundas is dry and the vegetation there is quite unlike that of the rest of Indonesia. The most characteristic tree is the Lontar Palm (*Borassus flabellifera*) whose stem is reasonably resistant to the frequent fires. This can be seen on the brown hills and against the skyline. The monsoon forest in the foreground of this view in Komodo National Park is made up of *Acacia* and other trees.

and livelihood of the human communities. Fires, deliberate or accidental, sweep through the lowland hills with the effect of increasing fodder production for cattle, goats and sheep. The remaining natural forests are under increasing threat because the inland people need to range widely to find sufficient fodder for their animals. While this did not matter when the population density was low, the rapid increase in the population and their herds is causing serious environmental degradation.

Sandalwood Trees

The Sandalwood tree (*Santalum album*), source of the valuable aromatic sandalwood oil, is a native of these islands. The tree has been utilized for over a thousand years and may have first been taken to India from here by traders in the fragrant wood. It is a medium-sized tree that grows in relatively dry savannah woodland. It is a parasite when young, growing on other plants and obtaining nutrients from them, and it appears that many plants act as host. The remaining natural stands of Sandalwood can be found on Sumba, Flores and Timor (particularly in the east), and many of these are exploited for the heartwood from which can be extracted the thick yellow oil used in incense, perfumes and various medicines. After the felling of large Sandalwood trees, replacement trees grow in two ways; first, from shoots growing from the old tree stump and, second, from seedlings. Unfortunately, both these shoots and seedlings are susceptible to fire and cattle grazing and, unless these and illegal logging are controlled, the Sandalwood forests will soon be lost.

Palms for Sugar and Paper

Perhaps the most common and characteristic trees of the drier areas are the Lontar or Palmyra Palm (*Borassus flabellifera*) and Gebang Palm (*Corypha utan*). The Lontar is found throughout the drier parts of South-east Asia as well as in Africa, India, Indo-China and Australia, and is conspicuous by its very dense crown of blue-green fan-shaped leaves. These leaves have a singular place in Indonesian history because the first books were written on them until the Portuguese brought paper. Nowadays the palm is used for a host of services, but the most economically important of these is as a source of sugar. This is obtained by cutting and bruising the enormous flower-stalk which grows from the centre of the palm, signifying the end of its life. The cut stalk exudes sugar which is collected in a length of bamboo. The liquid is then boiled slowly to drive off the water, leaving sticky brown sugar. The fruits of this palm are like small black coconuts and the flesh inside is sweet and pleasant to eat.

The Gebang Palm has only very small fruit but in all other ways it is much more massive than the Lontar. The leaf bases characteristically remain on the tree for some time, giving the trunk below the crown a rather rough appearance. The leaves themselves reach 2 metres (6 feet) across and the stalk has large dark spines. These leaves, too, were used for writing, but were considered superior to Lontar and reserved for the more important documents. Gebang flower-stalks are among the largest in the world, reaching about 5 metres (16 feet) high and 7–10 metres (23–33 feet) across with perhaps millions of flowers producing hundreds of thousands of fruit.

Animals

Like the flora, the fauna of the Lesser Sundas is also rather poor in species, and the trend is for gradual impoverishment towards

The Lontar Palm is frequently seen around villages in the Lesser Sundas, such as here in Flores, because it has multifarious uses – the leaves are made into water-carriers and other baskets, the trunk is used for firewood and timber, the fruit for eating and the production of an alcoholic beverage, and the sap for sugar.

the east. For example, among the toads and frogs, there are twenty-one species known from eastern Java, eleven on Bali, seven on Lombok and Flores, five on Timor and two on Alor. Among the birds, however, there is a very respectable list of 242 resident species, sixty-six of which are endemic. Animals which can fly or swim are present in greater variety because they can disperse between the islands more efficiently. There are very strong Australian affinities among the birds with parrots, cockatoos, honeyeaters and friarbirds being found throughout the islands. There are also eight endemic mammals comprising four fruit bats, two insectivorous bats, one shrew, and the Flores Giant Tree-rat (*Papagomys armandvillei*). Almost all the larger mammals found today on the Lesser Sundas, such as monkeys, civet cats, pigs and deer, were introduced by man many years ago. Exceptions to this are the two species of marsupial cuscus on Timor. The waters around these islands are also used as migration routes by various species of whales. The Lesser Sundas has seventeen endemic species of reptiles, including the rather rare dark-coloured Timor Python (*Python timoriensis*) which lives in the grasslands and open forests. But the area's most famous animal is surely the massive Komodo Dragon, the world's heaviest lizard.

Komodo Dragon

When Governor van Steyn van Hensbroek landed on Komodo in 1910 to obtain the very first specimens of this large lizard to be sent to a scientist, a couple of Dutch pearl fishermen there regaled him with stories of the 'lizards that had got away' – giant Dragons as much as 7 metres (over 25 feet) long. Stories of such outlandish beasts were perpetuated in the outside world, but the longest that a Komodo Dragon or Ora (*Varanus komodoensis*) is reliably reported to have reached is about 3 metres (10 feet). None has in fact been found this long for many years; more typical large specimens now are about 2.8 metres (9 feet) in total

The Komodo Dragon (*Varanus komodoensis*) is the world's heaviest lizard, reaching 50 kilograms (110 pounds) in weight and nearly 3 metres (10 feet) in length. Tourists come from all over the world to see this dramatic animal in its natural state.

length, weighing about 50 kilograms (110 pounds); still a pretty fair-sized monster, with males generally somewhat longer than females.

No other large carnivore has such a restricted distribution as the Komodo Dragon, which is known in historical times only from the western end and north coast of Flores, and from the small islands of Komodo, Padar and Rinca to the west. The small islands are today a part of the Komodo National Park and the Komodo Dragons there are reasonably secure, but the Flores populations are threatened by land use changes resulting from an expanding human population. Komodo Dragons occur from sea level to an altitude of 700 metres (2,300 feet) and may occur higher than this when travelling between areas. They can swim quite well and will enter salt and fresh water without encouragement. There is no fossil evidence that this species has ever occurred on neighbouring islands, but during periods of the Pleistocene, when the sea level was lower, the total land available to them was probably double what it is now.

The Komodo Dragon's habitat is shaped by the low, seasonal rainfall and the generally mountainous terrain. The typical vegetation is open savannah with Tamarind trees and Lontar Palms, and it is strongly influenced by fire, which keeps the habitat open and prevents scrub growth from taking over. Very frequent fires, however, cause a gradual degradation of the habitat. Komodo Dragons occupy burrows, hollow trees and other holes which tend to be cooler during the day and warmer at night than the outside air. Smaller individuals will use holes quite high up a tree and larger individuals use holes dug by porcupines, civet cats and rodents, but they dig most of their own burrows along dry creek beds and on open hillsides.

Komodo Dragons are most active during the day, although they may also roam abroad on moonlit nights. They can travel up to 10 kilometres (6 miles) in a day, though usually rather less, as they hunt in search of carrion and prey. When Komodo Dragons walk they swing their body wildly from side to side, but when they run the body is held stiffly and the sound of their feet hitting the ground is not unlike the sound of a muffled machine gun. They have been recorded sprinting at up to 18 kilometres (11 miles) per hour, a formidable foe indeed.

It is estimated that there may be as many as 7,000 Komodo Dragons, including some 1,500 hatchlings. They lead a rather solitary existence, but will gather at carrion, and males will fight for food and mates. While some of this fighting is largely symbolic, some is decidedly serious and can result in the death of the vanquished. The females lay an average of eighteen eggs over a period of days during the dry season, mainly in August and September. Newly hatched Komodo Dragons are colourfully speckled which affords them cryptic camouflage among the dense scrubby vegetation. The hatchling and smaller Komodo Dragons are mainly arboreal predators on insects and small lizards, but by the time they are about a year old they have graduated to birds and rats. When adult they are uniformly grey-coloured although they pick up the colour of the soil on their skin. This 'natural' colour may provide advantages when ambushing their prey. Juveniles are quite capable of climbing trees but, with increasing age and weight, the tail becomes less of a balancing aid for climbing and more a defensive weapon, and the thicker and heavy legs become more useful for holding down a carcass while tearing off flesh than for agile climbing.

One might think from watching Komodo Dragons in the national park that they are scavengers, feeding on rotting carcasses, and it has even been stated that they are unable to kill anything larger than a chicken. This is not supported by field observations, however, for there simply are not very many dead animals littering the islands except when natural events such as droughts or falls of volcanic ash occur. Most of the scavenging is done by the large adults, but they will also kill as and when the opportunity arises, and trials suggest that they would prefer to eat fresh rather than rotting meat. The prey of the larger Komodo Dragons is predominantly deer and pigs, but a wide variety of other animals are also taken including water buffalo, monkeys, snakes, palm civet, rats and goats. Their appetite is prodigious and a hungry Komodo Dragon is able to eat 80 per cent of its own body weight in food in a single day. Like many reptiles, the Komodo Dragons can also survive for long periods without food. Interestingly, almost all the major prey species have been introduced to the islands by man in historical times and it is not clear what Komodo Dragons ate in prehistoric times. Perhaps it was the small Pleistocene elephants which were found here.

It is very rare for humans to be attacked by Komodo Dragons but this does happen, generally when the animals are provoked. Near the observation area in the national park is a plaque which reminds visitors of the fate of a German tourist who approached too close to the Komodo Dragons and was killed before he could be pulled away. It is often stated that the bite of a Komodo Dragon is deadly even if the animal (or human) escapes. Tests have indeed shown that various unpleasant organisms can be found in the saliva, and festering sores are the almost inevitable result of an untreated bite. This is useful to the Komodo Dragon because it ensures that any deer or other prey which is wounded but escapes will probably die a few days later and can be tracked by its smell and eaten as surely as if it had been killed outright. The Komodo Dragon's sense of smell is very good and its bright yellow tongue can be seen flicking in and out of its mouth as it picks up air-borne chemical clues to the presence of food.

Whale Harvest
Off the eastern end of Flores lie two small islands called Lomblen and Solor situated in the middle of a major migration route for the great whales travelling to and from the rich southern oceans. Two small villages here are peculiar because their fishermen hunt not just for fish but also for whales.

Between May and August each year, the men of the villages take to sea in their long boats at sunrise in search of Sperm Whales (*Physeter catodon*) which come past in small groups of three to five. On sighting a whale the men lower the plaited-palm sail and row towards their quarry. As they close on the animal, the harpooner leans out from the bow of the boat and throws a homemade harpoon into the chosen whale. He will sometimes even get out of the boat and stand on the whale's back to ensure a good shot. After a long and dangerous chase, which may last a day or more, the exhausted whale dies and is towed back to shore amid great rejoicing. It is cut up and distributed according to tradition – nothing is wasted, meat, bone and oil all find a use. The coastal people barter these products with the inland farmers for rice and corn. Only the teeth of the whales are traded off the island. About twenty or thirty whales are taken in this way each year and so this traditional fishery represents no major threat to the wild population.

Two of the three different-coloured crater lakes in Kelimutu Volcano, eastern Flores. The range of colours is caused by different proportions of minerals in the surrounding rocks. This spectacular natural phenomenon attracts many domestic and foreign visitors.

Conservation

Plans are afoot from the Indonesian Directorate-General of Forest Protection and Nature Conservation and the World Wide Fund for Nature to give to the Lesser Sundas the conservation attention it deserves and which has been somewhat overdue. It is expected that professional and technical assistance will be given to help improve the management of existing conservation areas and to assess the conservation potential of other forested areas for protection, management and tourism development. An important component will be the involvement of local communities in the decision-making processes, in extension and educational programmes, and in the management and provision of visitor services. Of particular significance will be the identification of potential buffer zones around important conservation areas to provide resources for surrounding human populations while reducing the destructive exploitation of the region's remaining forests. Komodo National Park has already received considerable support and is a very popular location for tourism, and so special attention will now be given to Mount Mitis on west Timor, Wanggameti on Sumba and Mount Rinjani on Lombok, since they are all top priorities for conservation action and have been designated as areas of global conservation importance by the World Conservation Union (IUCN). All this will require a considerable investment of time and money but the potential results make this most worthwhile.

Irian Jaya

Irian Jaya is a land of vast tropical swamps, pathless forests and remote highlands. Even in these days of modern communications it must contain some of the wildest places left on earth. For many communities in the mountain plateaux, landing strips suitable for single-engined aircraft are the only means of contact with the outside world. Still other communities remain largely out of touch, and in the extensive swamplands travel is only possible by water. This, then, is the easternmost province of Indonesia, comprising the western half of the huge and exciting island of New Guinea and some of the islands to the west. New Guinea has the greatest contiguous area of rain forest outside Amazonia and a very rich range of habitats from quite unspoilt coral reefs to extensive mangrove swamps, inland swamps, lowland forest, montane forests, alpine meadows and, most surprising of all, glacier-topped peaks. In terms of biodiversity, Irian Jaya is comparable to Borneo, but whereas many Bornean species are found also on Sumatra, Java and Peninsular Malaysia, half of the plants and animals of Irian Jaya are not found outside the island of New Guinea.

Plants

The plants of Irian Jaya have affinities both with those of Asia and those of Australia. For example, in the lower montane forests, oaks typical of northern and western regions grow next to Antarctic beeches typical of Australian, New Zealand and South American forests. In the lowland forests, the eight dipterocarp species, so abundant in Sumatra and Borneo, can be found rubbing canopies with enormous Monkey-puzzle trees (*Araucaria cunninghamii*) also found in Queensland and Chile. Very few of the trees have a high economic value at present and those that have, such as Black 'Walnut' (*Dracontomelum mangiferum*) and ebonies *Diospyros*, are highly dispersed. Most of the Irian Jaya timber trees such as *Intsia bijuga*, *Intsia palembanica* and *Pometia pinnata* are of decidedly secondary importance elsewhere in the archipelago.

Among the probable total of 16,000 plant species (about one-third of them orchids) there are at least 124 endemic genera (compared with fifty-nine on Borneo and ten on Java). One of the best known of the plants is the cultivated Flame of Irian (*Mucuna novaeguineensis*) whose large, bright red, sweetpea-like flowers hang in dense garlands below the climbing stems. Hundreds of different plants and plant parts are distinguished by the indigenous peoples of Irian Jaya because they are used as medicines, narcotics, stimulants, tools, weapons, construction materials and cloth, and for ceremonial dress as well as everyday ornamentation.

Swamps

The most extensive swamps in South-east Asia are found along the southern coast of the island of New Guinea, and most of them lie in Irian Jaya. There are also major swamps along the north coast around the Mamberamo delta and in its headwaters along the Tariku and Taritatu rivers. Here one finds plants which are restricted to wet habitats and also those which put up with being inundated for part of the year. Most characteristic in the seasonally inundated areas are the fire-resistant Paperbark

A sluggish river meanders through extensive lowland forest in Irian Jaya; there is more contiguous rain forest in Irian Jaya than anywhere else outside Amazonia.

There are a staggering 157 species of *Rhododendron* on the island of New Guinea, only two of which are found elsewhere. Most of them grow in the mountains, as here in the Baliem Valley of the Central Highlands of Irian Jaya.

(*Melaleuca*) woodlands of south-east Irian Jaya which lie in the monsoonal rainshadow where they are sometimes the only tree species present. The Screwpines (*Pandanus*) are frequently seen in swampy areas, and the Sago Palm (*Metroxylon sagu*) is also very common and can form extensive woodlands in shallow swamps. Sago flour is the staple of all the lowland peoples. It is obtained by felling a palm and grating or chopping its pith; this is then soaked and strained to wash out the starch which is cooked as required. This process is wonderfully efficient, providing people with ample time to fish and hunt. The Sago Palm also provides firewood, roofing material, fish bait, pig food and many other necessities.

Mangroves

Mangrove forests are typical of tropical coastlines, and throughout the area from India to the Pacific their composition is broadly similar with only about twenty tree species commonly encountered in the regularly inundated areas. Further inland more species are encountered. Since each species has its own set of requirements and preferences, the different species tend to be found in distinct zones through the forest. Mangrove trees are

The Arfak Reserve near Manokwari, on the north-east of the Bird's Head Peninsula, is a rugged area of steep mountains and wonderful forest. Half of all the bird species found in Irian Jaya are present within the reserve, and half of these are endemic to the region.

characterized by their tolerance of saline soils – these may not, in fact, be their preferred habitat, since some grow perfectly well in fresh water, but there is too much competition from other plants in freshwater habitats to enable them to become established. Mangrove forests are unusual among tropical forests in that there are relatively few shrubs and climbing plants, because of the regular tidal flooding of the forest floor. Irian Jaya has some of the most extensive mangrove forests left in South-east Asia. Elsewhere in Indonesia, where human population densities are much higher, many areas of mangroves have been converted into brackish water fish and prawn ponds, or felled for timber or to make rayon fibre. The influence of these activities in Irian Jaya is as yet insignificant in comparison with the areas of virtually pristine mangrove forest.

When soil is inundated for long periods, the air spaces within it become filled with water and the soil loses its oxygen. These

Part of the vast swamp forests in southern Irian Jaya near Timika. The palm trees are Sago Palms (*Metroxylon sagu*) and the leafless branched structures are their flowering stems. It takes ten to fifteen years for a Sago Palm to build up enough reserves to flower and, having done so, it dies.

conditions would kill most tree species, as can be seen in some logging areas where roads have inadvertently dammed small rivers, killing the trees behind them. The trees in mangrove and other swamps are able to cope in a number of ways. Some are able to respire without free oxygen whereas others have very large pores or lenticels in their bark which allow oxygen to reach the growing tissues, and still others have various forms of root which grow upwards and emerge above the soil surface.

Perhaps the best-known features of mangrove trees are their peculiar aerial root systems. Three sorts can be found: the stilt roots of *Rhizophora* trees which seem to walk out from the trunk and may prevent other trees from growing too close; the undulating knee-roots of *Bruguiera*; and the spike-like root extensions of *Sonneratia*. One of the more common trees, *Ceriops*, does not have unusual roots, but the tree trunk has exceptionally large pores in it to allow for gas exchange.

Rhizophora flowers are probably wind-pollinated, but many of the other species engage the services of animals for pollination. *Sonneratia*, for example, is pollinated by bats which may fly as much as 40 kilometres (25 miles) from inland caves each night to feed, while the anthers in the flowers of *Bruguiera* and *Ceriops* open explosively when the flowers are visited by curved-beaked sunbirds or moths.

As a rule, fruit develop on a plant and when they are ripe and the seeds ready to germinate, they are released from the parent. Interestingly, *Rhizophora* and *Bruguiera* mangrove trees are among the few plants which differ by having fruit which actually germinate on the parent before being released. The thick root grows down from the fruit and may be 45 centimetres (18 inches) long before it drops like a lance into the mud or water below.

The Dry South-east

Although one thinks of Indonesia as being forested with dripping, dark green vegetation, down in the south-east corner of Irian Jaya it is more reminiscent of parts of Australia: there are vast expanses of grasses with only isolated trees. This area has a very marked dry season and is one of the driest parts of Indonesia. As a result, dry savannah forests have developed. The trees growing here are very different from those growing in the rain forests, with species of *Melaleuca, Eucalyptus, Casuarina*

and *Acacia* predominating. The composition of the woodlands is much influenced by fires which are set regularly by local inhabitants to improve conditions for hunting. Some areas are taken over by a fire-induced and fire-maintained vegetation of bamboos, particularly of the genus *Schizostachyum*. Typically here one finds the Agile Wallabies (*Macropus agilis*), which occupy a deer-like niche in the ecosystem, and introduced deer, as well as hosts of waterbirds including pelican, egrets, ducks and ibises, cranes and the tall, strutting, proud-looking Bustards (*Ardeotis australis*).

Mountains

In the high mountains and plateaux above 3,000 metres (10,000 feet) in the central range the vegetation is strikingly different from that in the lowlands. Some of the conifer trees of lower altitudes are still present but they are more crooked and have dense growths of lichens hanging from them. They are interspersed with tree ferns (*Cyathea*), bogs and grasslands, reminiscent of more temperate latitudes. A wide variety of *Rhododendron* species, many with exciting, potentially prize-winning blooms, can be found here together with many bilberries and blueberries (*Vaccinium*) and nearly a hundred species of orchid. The mammal fauna is not rich, comprising mainly hardy rats and dwarf wallabies. In the harsh environment above about 4,200 metres (13,800 feet) the vegetation comprises low tussocks of sedges and grasses, clumps of *Rhododendron* and *Vaccinium*, patches of lichens, and small herbs such as species of buttercups (*Ranunculus*), gentians (*Gentiana*), cinquefoils (*Potentilla*) and many others. Among the birds restricted to these elfin and grassy areas are the speckled Snow Mountain Quail (*Anurophasis monorthonyx*) and the white-throated Short-bearded Honeyeater (*Melidectes nouhuysi*).

Mountains warm up and cool down much more quickly than the lowlands, primarily because the air above them is less dense. Daily temperature ranges at high altitude can be as much as 15°–20 °C (60°–70°F). Frosts occur above about 3,500 metres (11,500 feet) and ultra-violet radiation is not absorbed so bad cases of sunburn can occur if precautions are not taken. It has been suggested that the high ultra-violet radiation received on tropical mountain tops may result in increased rates of cell mutation, causing the high levels of endemism found here.

Clouds form when the air reaches the temperature at which the water vapour condenses out and when there are dust particles on which the condensation can occur. During wet periods when the water vapour content of the air is high, the whole of a mountain can be swathed in cloud, but at drier times it is more usual for a belt of cloud to form, often at about 2,500 metres (8,200 feet) leaving both the higher peaks and the lower slopes clear. There are no clear patterns associated with rainfall on mountains, except that rainfall on the slopes up to about 1,500–2,000 metres (5,000–6,500 feet) is more than on the surrounding lowlands and is generally greater on the side facing the prevailing wind.

Mammals

The geological history of New Guinea has been the major determinant of its mammal fauna. Of the 174 non-marine mammals known from Irian Jaya, 100 are endemic species, half of them rats and bats originating in Asia, and half marsupials originating in Australia. The marsupials are those animals which give birth to very undeveloped young which then crawl into a pouch on the mother's belly, where they attach themselves to a nipple until large enough to face the outside world. These interesting beasts include some strange endemic genera of carnivorous marsupial mice of the family Dasyuridae (the family that includes the possibly extinct Tasmanian Wolf), and the fecund bandicoots of the family Peramelidae whose habits resemble those of shrews and hedgehogs. The marsupial mice are additionally interesting in that their pouches open towards the back, so that they do not fill with earth while the mice are burrowing or moving around underground. Their gestation period is the shortest of any mammal, about thirteen days. There are eight species of arboreal cuscus, each with naked skin beneath their prehensile tails, and thirteen species of possum including the Sugar Glider (*Petaurus breviceps*) which can glide for up to 50 metres (165 feet) between trees, using the flaps of

skin between its arms and legs, in a manner similar to flying squirrels. Among the rather unglamorous rodents in Irian Jaya there are the Rough-tailed Giant Rat (*Hyomys goliath*) and Smooth-tailed Giant Rat (*Mallomys rothschildi*), and the tiny New Guinea Jumping Mouse (*Lorentzimys nouhuysi*). The largest terrestrial mammal is the Rusa Deer (*Cervus timorensis*) which was introduced in historic times, and has now spread widely through the province. There are also four tree kangaroos (*Dendrolagus*) and five forest wallabies (*Dorcopsis*) in the forests, and plains wallabies including the Agile Wallaby (*Macropus agilis*) of the dry south-eastern plains and woodlands. Although many of these marsupials are perhaps strange to our eyes, it is possible to recognize them ecologically; that is, in the near absence of Asian mammals, some of the Irian species take on the role of shrews, others of primates, others of small carnivores, and still others of deer.

The most typically Australian mammals in Irian Jaya are the kangaroos, wallabies and echidnas, and these are described in more detail below.

Kangaroos and Wallabies

It is perhaps surprising to see several species of kangaroo-like animals in the forests and savannah plains of Irian Jaya, taking the ecological role of primates and deer. The long hind-legs of the bush wallabies allow them to travel at speeds up to 40 kilometres (25 miles) per hour, with their heavy tails providing balance. Although protected, they are hunted both by traditional hunters and by townsfolk from Merauke, with damaging effects. The tree kangaroos have a different body shape from their relatives with their four legs being roughly the same length, and their feet and hands broader and rougher than those of their earth-bound cousins. They are accomplished climbers and have large, strong claws with which to grip onto branches. They can also jump as far as 6 metres (20 feet) between trees, their tails acting as rudders. In contrast, they are peculiarly gauche if they have to walk on the ground. Tree kangaroos are

The Spotted Cuscus (*Spilocuscus maculatus*) is a very widespread species found throughout New Guinea, west to Seram and Buru, and south to the northern tip of Australia. It is entirely arboreal and nocturnal, and eats leaves and fruit.

An undescribed species of tree-kangaroo (*Dendrolagus* sp.) being held by a tribesman near Tembagapura, below Puncak Jaya. Tree-kangaroos are relatively easy prey for hunters, and they have become very shy as a result.

rarely seen in the wild, partly because they are well camouflaged against the forest and partly because they are much sought after for food by the local people and so have become very secretive. It is thus very difficult to positively determine their conservation status, but it is probable that their numbers have seriously diminished in many areas.

Echidnas

Certainly the most unlikely mammals in Irian Jaya are the two species of echidnas or spiny anteaters. These, together with the Australian Platypus, form the mammal group called the monotremes, distinguished by their egg-laying habit. The Short-beaked Echidna (*Tachyglossus aculeatus*) has a genteel if somewhat long snout, and dark-tipped spines all over its body protruding through the fur on its sides and back. The Long-beaked Echidna (*Zaglossus bruijni*) has a very long, downward-curving snout accounting for one-quarter of the animal's total length, and relatively small and few spines which protrude only through the fur over the neck, elbows and knees. The latter species is thought to be under some threat from hunting by natives, but it is extremely difficult to assess its status accurately.

The spines serve the same purpose as in the hedgehog – when an echidna is threatened it rolls into a ball making a singularly unappetizing dish for a predator such as an eagle or monitor lizard. If it is on soft soil it can use its remarkable claws to burrow straight downwards, leaving just the tips of its spines above the surface. It can also use its claws and spines to wedge itself very securely in rock crevices. Echidnas detect danger by hearing rather than by sight, for their eyes are of little use, but they can generally hear an approaching human long before it can be seen. Food is detected by smell while foraging among the leaf litter or undergrowth. The Short-beaked Echidna uses its long sticky tongue to lick up ants and termites, whereas the Long-beaked feeds mainly on earthworms which it hooks with spines in a groove in the front of its tongue. The worms have to be taken into the narrow beak either head or tail first, and the echidna uses its forefeet to carefully position them.

Echidnas lead largely solitary lives occupying overlapping home ranges in wet forest. Within these they have no regular resting or sleeping areas, merely taking cover where and when they can. The only fixed point is the female's nest burrow where she lays her single soft-shelled egg. This hatches after about ten days and the young crawls into the mother's pouch where it is nourished for about six months from milk-producing pits rather than nipples.

Birds

To date, 643 species of birds have been recorded in Irian Jaya, including 269 species endemic to the island of New Guinea. No other island or major island group in the country can match this level of distinctiveness. Most of these birds have affinities with the Australian avifauna, but Blyth's Hornbill (*Rhyticeros plicatus*), the tree swifts (*Hemiprocne*) and the shrikes (*Lanius*) are decidedly Asian and reach their easternmost limit in New Guinea. As well as the glorious birds of paradise and bowerbirds, there is the world's most diverse assemblage of kingfishers, many of which neither live near water nor catch fish, and a large number of honeyeaters. There are also forty-two species of parrot in the province, from the stocky and long-necked Vulturine Parrot (*Psittrichas fulgidus*) with its black body and red belly, rump and wing patch, to the minuscule Pygmy Lorikeet (*Charmosyna wilhelminae*), just a scrap of darting green

A Victoria Crowned Pigeon (*Goura victoria*) and its chick. This species, identified by the white tips of the crown, is found in northern New Guinea and is one of three similar species found in Irian Jaya.

with flashes of bright red beneath the wings of the male. There are many pigeons but the most impressive are the three species of geographically separated, turkey-sized Crowned Pigeons (*Goura*), the largest pigeons in the world, all of which are mainly steel-blue with a Mohican-type crown of lacy feathers. They live in small flocks in flat lowland forests and are rather tame. When they do take off from the ground their wings slap noisily together as they try to gain height. They are said to be found at the sites where people process sago flour, possibly attracted by the insects which also seek out this carbohydrate-rich food source. Another extraordinary ground-dwelling endemic pigeon is the Pheasant Pigeon (*Otidiphaps nobilis*) which is about the size of a small chicken but with a small crest at the back of the head, a V-shaped tail, rounded wings, and long legs which give it a superficial resemblance to a pheasant. Its feathers are iridescent blacks, greens and purples, and the small wings are a chestnut brown. These birds are not particularly rare but they are shy, taking to the wing for short flights only when they have to, and are seen only singly or in pairs. The largest predators on the island are birds: the massive New Guinea Harpy Eagle (*Harpyopsis novaeguineae*) and the Wedge-tailed Eagle (*Aquila audax*).

Of all the birds of Irian Jaya the most exciting are the raucous Palm Cockatoo (*Probosciger aterrimus*), the exquisite birds of paradise, the industrious bowerbirds and the enormous cassowary, and these are described in more detail below.

Palm Cockatoo

The fearfully loud braying of the large Palm Cockatoo is one of the more unlikely sounds in the Irian forest. This slow-flying, black-crested parrot with conspicuous red cheek-patches also has a large repertoire of musical but equally loud notes, some sounding like the squeals from a gigantic guinea pig. It possesses a massive black beak with which it opens large, hard

The Sulphur-crested Cockatoo (*Cacatua galerita*) is found throughout the island of New Guinea and in much of Australia. In Irian Jaya it is locally common in lowland forests, although it is scarce in more accessible areas where it is hunted for tribal decorations and because it raids crops, as well as being trapped for the caged bird trade.

seeds such as those of *Terminalia*, kenari, pandans and palms. It also eats berries, soft fruit and leaf buds, and uses its beak to extract soft palm shoots by splitting apart the leaf fronds. The long, rounded wings carry it across the tree tops in a straight, level flight characterized by several slow flaps then a short glide. Palm Cockatoos are mainly seen singly or in pairs, and occasionally in small groups at a good food source. They nest in hollow trees where a single egg is laid; it is not known how long they live in the wild, but captive specimens have lived thirty-five years. As are most of the parrots in Irian Jaya, it is protected by law but some are still caught and smuggled out of the province to satisfy the demands of cagebird enthusiasts in the west. The species is beginning to be bred in captivity and young birds can command very high prices.

Birds of Paradise

Among the most extravagant of the trade goods brought from the eastern islands hundreds of years ago were the flamboyant feathers of the birds of paradise. These were worn by the bodyguards of Turkish sultans 600 years ago, as well as in other royal courts in succeeding years. The term 'birds of paradise' was given for the straightforward reason that it was believed until well into the seventeenth century that they did indeed come from Paradise. Their extravagant colours were thought to be the result of flying close to the sun and it was accepted that they never alighted on earth until they fell to the ground at the end of their given span. This belief was reinforced by the lack of feet on the skins of birds brought to Europe. Such birds clearly did not need feet, and the father of modern taxonomy, Carl Linné, named the largest of the species, the Greater Bird of Paradise, *Paradisaea apoda*, the 'legless bird of paradise'. The reason for this curious state of affairs was that the native hunters who used the feathers themselves in their ceremonial regalia, traditionally removed the feet when they skinned their quarry.

Alfred Russel Wallace spent eight months on and around Irian Jaya in 1858–60, and he referred to these birds as 'the most extraordinary and the most beautiful of the feathered inhabitants of the earth'. This and other unsolicited testimonials encouraged trade in the feathers, which really hotted up when Parisian designers decided that the plumes set off the curves and colours of their clients. As a result, thousands of birds were killed to be plucked or, more grotesquely, stuffed and mounted atop hats. This trade was banned in the 1920s, but human curiosity and status-seeking have encouraged the continued capture and sale of these glorious creatures. Even today there is an active smuggling network and some forest areas are virtually devoid of these birds.

There are about twenty-six species of birds of paradise in Irian Jaya, but only two-thirds of these are the spectacular specimens which have made the family so famous and prized. These are the polygamous species such as the Twelve-wired Bird of Paradise (*Seleucidis melanoleuca*) and King of Saxony Bird of Paradise (*Pteridophora alberti*), the males of which develop their most dazzling feathers in the breeding season when they display for the pleasure and enticement of the singularly dull females. Some of the species, such as the Greater Bird of Paradise, perform these displays at special tree-top perches or

A female Twelve-wired Bird of Paradise (*Seleucidis melanoleuca*), a bird of the seasonal swamps, particularly near Sago Palms and pandans. It feeds on fruit, arthropods and perhaps nectar, often joining flocks of other species. The black and yellow male has twelve very fine long feathers curling round from its tail which it flaunts during courtship displays.

'leks' where many males will come together to try to outshine one another to 'win' a female. After mating the female is left on her own to build a nest and rear their young, while the male returns to the lekking tree.

The monogamous birds of paradise have relatively dull plumage. An example of this is MacGregor's Bird of Paradise (*Macgregoria pulchra*), a black-coloured inhabitant of the highest mountains on the island, distinguished by large yellow wattles around the eyes and striking yellow wing-patches. These birds mate for life, so the males do not need to display regularly to attract a mate, and they share in the building of the nest and the rearing of the young.

Bowerbirds

Close relatives of the birds of paradise are the bowerbirds which are found in Australia and New Guinea. Irian Jaya has some eleven species, mostly found in the mountains, ranging in size from thrush- to crow-size. Like the birds of paradise, the species fit neatly into two groups: the relatively dull, monogamous species, and the polygamous species with rather drab females and flamboyantly adorned males, many with fiery orange crests. Although they spend most of their lives in the trees, the habit for which the polygamous species are best known occurs on the ground, for this is where the males build their elaborate bowers. These are built with sticks and some species also ornament their constructions with carefully chosen objects of a certain colour, or with fruit pulp, chewed grass or charcoal. The builder of the largest and most elaborate bower in Irian Jaya is the Bird's Head Bowerbird (*Amblyornis inornatus*) which constructs a cone-shaped hut up to 1 metre (3 feet) high and 1.6 metres (5 feet) across, with a door on the entrance and a front 'lawn', both of which are decorated with flowers and fruit. Some of the species, such as the Grey or Savannah Bowerbirds (*Chlamydera*) and the Golden Bowerbird (*Sericulus aureus*) build bowers called 'avenues' because they comprise a pair of high fences built so strongly with interwoven twigs and grasses that they can be lifted off the ground. The male brings display objects such as stones to place in the centre and at the openings of the bower to entice the female into the middle. Each morning the male fusses around his bower, smartening up the twigs and trinket collection until a female comes in view. He then begins exuberant dances to attract her, accompanying himself with whistles, hisses and clicks. After a successful mating, he will abdicate responsibility for chick rearing, leaving all the nest-building, incubating and feeding of the young to the female.

Cassowary

You are unlikely to see cassowaries in the forest, partly because they are wary of man and partly because they are mainly solitary and nocturnal, but it is quite likely that you will hear their deep booming or croaking calls. Also, by keeping your eyes open along river banks and along forest trails you may see the dinosaur-like foot-prints in the mud, or piles of droppings containing masses of seeds. The cassowaries' wariness is fortunate because they are arguably the most dangerous New Guinea animal. Of their three big toes, the innermost one has a very long and heavy claw which is used when fighting other cassowaries. It is also not unusual, however, for humans to be attacked should the birds find themselves cornered or otherwise affronted, and severe injuries may result. Cassowaries are not beyond domestication, however; young ones are sometimes kept in villages to be fattened up for a feast and tame, full-sized cassowaries are sometimes allowed to roam in the scrub and forest around villages.

The Two-wattled Cassowary (*Casuarius casuarius*) lives in the lowland forests of New Guinea. Cassowaries feed primarily on fallen fruit, but may also take insects, small animals and some plant material. The female lays three to four eggs on the bare ground and the male looks after the chicks until they are independent. The cassowary's inner toe has a very long claw which is used in fighting and can cause serious injury. Young cassowaries are sometimes kept in villages to be fattened up for a feast.

These black flightless birds are by far the largest land animals on the island of New Guinea. All three species are found in Irian Jaya: the Southern or Two-wattled Cassowary (*Casuarius casuarius*), the Northern or One-wattled Cassowary (*Casuarius unappendiculatus*), and the Dwarf or Bennett's Cassowary (*Casuarius bennetti*). The first two grow to 1.5 metres (5 feet) tall and are found in the lowlands, whereas the Dwarf Cassowary reaches only just over 1 metre (3 feet) and is generally found in the hills and mountains up to the limit of tree growth. The cassowary plumage resembles coarse fur, as the narrow black feathers are flexible and droop. The wing feathers have no vanes and are reduced to a long black quill. A heavy helmet adorns the otherwise naked, bright-blue head, and the neck of the two larger species has a variety of red and yellow patches and wattles.

Cassowaries feed primarily on fallen fruit, mainly those that are red, black and orangey-yellow, but may also take insects, other small animals and some plant material. Some species of tree have fruit which are dispersed only by cassowaries, thereby forging a relationship of mutual dependence. The female cassowary lays three to four eggs on the bare ground and then ups and leaves the smaller male to look after them and the resultant active young until they are independent. The stripy chicks reach adult size in just one year, but have dull brown plumage for a further year.

Reptiles and Amphibians

The island of New Guinea is rich in reptiles and amphibians; allowing for the probability that not all the species have yet been discovered, it is likely that more than a hundred species of snakes and 200 species of lizards, many of them endemic, could be found there. In addition, there are two species of crocodile, six freshwater turtles, six marine turtles, and probably well over 200 species of frogs and toads most of which are descended from Asian forebears which made eastern peregrinations.

The vast majority of Irian Jaya's frogs and toads are found in the lowlands, but some have been found up to nearly 4,000 metres (13,120 feet). With over 200 species it is not surprising that they adopt a wealth of life strategies. For example, there are burrowing species, strictly aquatic species and terrestrial species, and some whose eggs develop directly into miniature forms of the adult, such as the nine endemic *Xenobatrachus* toads. The largest frog in Irian Jaya is the Arfak Frog (*Rana arfaki*) which reaches 16 centimetres (6 inches) in body length.

One of the most outstandingly beautiful snakes in the world is the harmless New Guinea Green Tree Python (*Chondropython viridis*), whose bright green skin is the colour of fields of young rice sprinkled with small yellow jewels. Strangely, the young are a pink, red, brown, or yellow colour. This species is very popular with hobbyists and zoos and may be threatened by the trade. It is, however, beginning to be bred in captivity and it is hoped that the demand for pets can be met from captive-bred specimens. Irian Jaya also has two of the world's most deadly snakes, the Death Adder (*Acanthophis antarcticus*) which is also found in Australia and west to Seram in the Moluccas, and the Taipan (*Oxyuranus scutellatus*) from the southern half of the island, which is also found in Australia. The Death Adder is extremely aggressive, making little attempt to escape when threatened or disturbed, and despite its relatively small size it can deliver a fatal bite to humans. In contrast, the Taipan tends to slither into holes when approached, but if cornered it will strike and bite, delivering a fatal dose of venom.

The largest, best known and possibly the most-feared reptiles in Irian Jaya are the crocodiles.

Crocodiles

It was reported in 1970 that a 7-metre (23-foot) man-eating Estuarine Crocodile (*Crocodylus porosus*) was finally killed in the Asmat area in southern Irian Jaya, but only after it had claimed an astounding fifty-five lives. Irian Jaya is among the major strongholds for this species but even here it is frequently hunted, with the result that numbers have declined significantly. There is also a less well known species called the New Guinea Crocodile (*Crocodylus novaeguineae*) which lives primarily in the Mamberamo river system in northern Irian Jaya.

Crocodile skins are a valuable commodity because of a

Adolescent Estuarine Crocodiles (*Crocodylus porosus*) held in a farm near Jayapura to supply skins for the leather trade. They were caught when small and grown in captivity. Regulations now require that a proportion of the animals are released back into the wild to produce the next generation.

regrettable consumer demand for handbags and shoes. The trade in wild crocodile skins is centuries old but, when it was realized that the wild stocks had been severely reduced, it was clear that the trade could not continue on that basis. One response to this from traders was to set up 'farms' to which wild-caught young animals were brought, grown on, and then skinned when they had reached an economic size. This approach exploited the village people who supplied the young crocodiles to the farms because they saw only a paltry proportion of the enormous profits that were being made. In recent years, moves have been made to develop a 'plasma' of village-level enterprises around a 'nucleus' farm with skinning facilities. This aims to ensure that the villagers see a fair return for their efforts, that the breeding stocks are protected, that only animals of a certain size are collected for the farms, and that a percentage of the animals of skinning size are returned to the wild to become the breeding stock of the future. This approach of giving the species a monetary value by its utilization should be a positive conservation measure just so long as people play by the rules.

Fishes

Irian Jaya has relatively few species of freshwater fishes, lacking any carps, minnows, loaches and other fishes typical of western Indonesia, but what it lacks in numbers it makes up for in interest. Probably the most dramatic of the Irian Jaya fishes is the Small-toothed Saw-fish (*Pristiopsis leichardti*) which is found in large rivers and lakes in the north and south of the lowlands. This is a shark-like fish which has a long, flattened 'beak' with a row of teeth jutting out from each side like a double-edged saw, and it can reach the impressive size of 5.2 metres (17 feet). It feeds by shaking its head rapidly from side to side among a shoal of fish and then eating the dead and injured ones.

More graceful and aesthetically pleasing is the Irian Bony-tongue (*Scleropages jardinii*). This is a relative of the species found in Sumatra and Kalimantan and the one in Australia, but has smaller barbels and delicate light spots and stripes on the scales and head.

Much more familiar to amateur aquarists are the rainbow-fishes, a family found in New Guinea as well as north and east Australia. These small colourful fishes with two dorsal fins are widely distributed in rivers, lakes, ponds and swamps. The tall mountains of the central range of New Guinea have separated two groups of species, and only one species from mountain streams is found on both sides of the range. New species are being found almost with every exploration of new areas, and many of them appear to be restricted to quite small areas. For example, the reddish-brown-striped Ajamaru Rainbowfish (*Melanotaenia ajamaruensis*) and purplish-brown Boeseman's Rainbowfish (*M. boesemani*) are known only from the Ajamaru Lakes in the middle of the Bird's Head Peninsula at the west of the province, and the dark *Glossolepis incisus* only from Lake Sentani, near the main Jayapura airport. Perhaps the most beautiful of the family is the diminutive Threadfin Rainbowfish (*Iriatherina werneri*) which has long, flowing and colourful fins which look like the product of years of careful selective breeding (as is the case with the common guppy). This fish was first described only in 1974 from specimens found in irrigation canals near Merauke in the extreme south-east of the province, although it has now been found on the northern tip of the Cape York Peninsula in northern Australia. Threadfins are remarkably hardy and are now bred successfully by aquarists for enthusiasts who thrill to the shimmering colours.

Insects

Although no one knows for sure, it is estimated that there are perhaps 100,000 species of insects on the island of New Guinea. Most are small and inconspicuous, but some of the beetles, stick-insects, katydids and cicadas can surprise and impress even the casual observer. The insect group that is most often noticed, however, is the butterflies.

Birdwing Butterflies
New Guinea has at least 5,000 species of butterflies and moths, among which the most spectacular must surely be the birdwing butterflies from the family Ornithopteridae, the largest butterflies in South-east Asia with wing-spans of up to 33 centimetres (13 inches). They are primarily inhabitants of the forest canopy and consequently relatively rarely seen or collected, and these factors, together with their size and beauty, have resulted in certain species being very valuable to collectors, with prices of thousands of pounds per specimen. The centre of diversity for this Asian group is in the Arfak mountains on the east of the Bird's Head Peninsula at the western tip of Irian Jaya.

It is possible to exploit these spectacular butterflies and other large invertebrates with commercial value in such a way as to benefit rural communities. It is hoped that steps in Irian Jaya will follow those used so successfully in Papua New Guinea in teaching villagers how to farm insects by growing, close to houses or in gardens, the specific plants on which the females lay their eggs. Harvesting and packing techniques for when the butterflies emerge from the chrysalis also have to be taught, and a transport system set up. The export of preserved butterflies could be co-ordinated from regional centres. Initially, just the insects would be exported from the province, but later products such as paperweights and framed collections could be produced. This type of enterprise would also help to eliminate the shadowy illegal export of these beautiful creatures.

Conservation

The challenges facing conservation in Irian Jaya, and the wonderful opportunities it offers, are very different from those in the more densely populated and developed islands of Indonesia. Large tracts of natural vegetation and wildlife remain amid a growing awareness that the indigenous people's need for development must be planned within a framework of the sustainable use of natural resources. The World Wide Fund for Nature in Indonesia has had a special conservation programme in Irian Jaya for over ten years which has sought to identify the important conservation areas, lobby for their protection, assist the government agencies to manage the areas, work on the ground with local people to effect boundary protection and sustainable patterns of exploitation, and to conduct biological surveys. The work continues.

Aerial view of Lake Sentani near Jayapura in north-east Irian Jaya. Among the inhabitants of its waters are 5-metre (16-foot) sawfish and exquisitely coloured rainbowfish found nowhere else.

Focus on Sumatra

Sumatra can be divided roughly into four sections: the broad eastern plains and swamps; the main range of mountains; the narrow western plains; and the chain of islands off the west coast. The long ranges of mountains forming the backbone of Sumatra were largely formed by compression and folding, but subsequent faulting resulted in the formation of a number of volcanoes such as Mount Kerinci, at 3,805 metres (12,483 feet) the tallest mountain in Indonesia outside Irian Jaya and the second tallest mountain in South-east Asia after Mount Kinabalu (4,101 metres or 13,455 feet) in Sabah (northern Borneo). Other famous Sumatran volcanoes include Merapi (2,891 metres or 9,485 feet) and Singgalang (2,877 metres or 9,439 feet) outside Bukittinggi in West Sumatra, and the classic Mount Sinabung (2,451 metres or 8,041 feet) north of Lake Toba. The largest complex of mountains in Sumatra are those surrounding the gigantic crater lake of Lake Toba, and those in Mount Leuser National Park in which can be found the second highest mountain in Sumatra, the non-volcanic Mount Leuser (3,381 metres or 11,092 feet). A number of large rivers drain eastwards from the mountains, such as the Rokan, Kampar and Indragiri in the centre, and the Musi towards the south. The sediment from these forms both deltas and offshore islands and the eastern coastline is continually, if slowly, changing.

Sumatra and its off-lying islands have a total area of 473,606 square kilometres (182,859 square miles), one quarter of the land area of the entire country, and are divided into eight provinces. In the north there is gas- and forest-rich Aceh (55,392 square kilometres or 21,387 square miles) which includes the western island of Simeulue, and North Sumatra (70,787 square kilometres or 27,331 square miles) with the world-famous Lake Toba, and the large western islands of Nias and Banyak. In the centre of the main island there are three provinces: oil-rich Riau (94,562 square kilometres or 36,510 square miles) which includes the many small islands off its eastern shore north-east to Natuna Island; narrow and mountainous West Sumatra (49,778 square kilometres or 19,219 square miles) with the Mentawai Islands; and forested Jambi (44,924 square kilometres or 17,345 square miles). In the south the three contrasting provinces are the large, resource-rich and industrial South Sumatra (103,688 square kilometres or 40,034 square miles) with the two large eastern islands of Bangka and Belitung; mountainous and forested Bengkulu (21,168 square kilometres or 8,173 square miles) with the remote offshore island of Enggano; and deforested and eroding Lampung (33,307 square kilometres or 12,860 square miles) including the famous Krakatau volcano in the Sunda Strait.

The population of Sumatra is 38.5 million and is growing faster than in other parts of Indonesia, by over 3 per cent per annum, at least in part because of the large immigration encouraged by the official transmigration programme and partly because of the

Paddy fields above Lake Maninjau, a crater lake in West Sumatra.

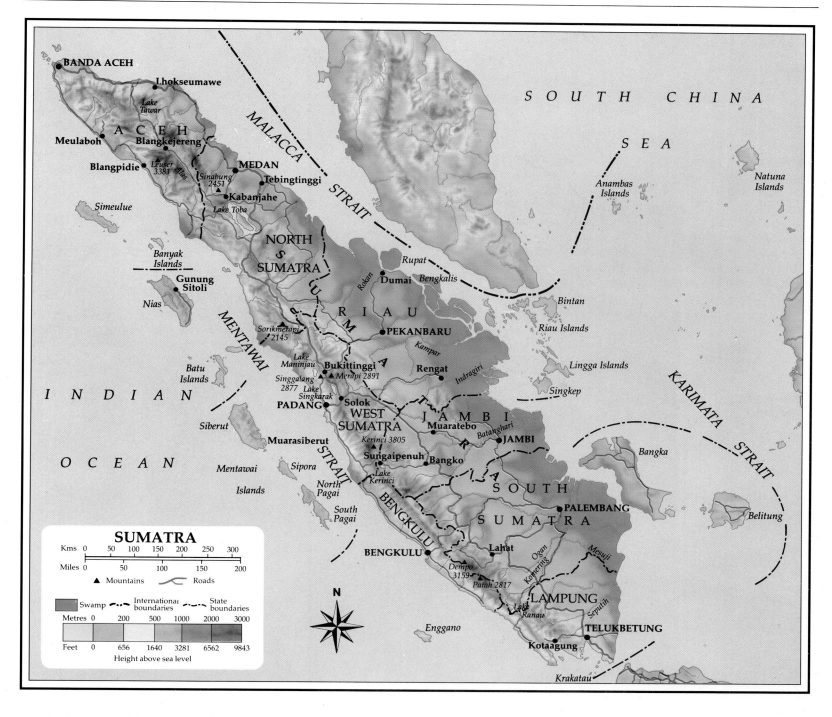

perceived opportunities that the island offers. Predominant among its peoples are the Bataks whose native lands are centred around Lake Toba. Other major groups include the proud Acehnese whose language is quite unlike any other, the matrilineal Minangkabau of West Sumatra, the tattooed inhabitants of Siberut Island, the Melayu of the east and south, and the Rejang around Lake Kerinci.

Almost all of Sumatra is more or less wet throughout the year with three months or less in which there is under 100 millimetres (4 inches) of rain. The average annual rainfall is 2,821 millimetres (111 inches), but some areas along the mountain chain receive nearly 5,000 millimetres (200 inches) in a year.

Just under half of Sumatra is covered with some form of forest but much of the lowland forest has already been logged or is due to be logged in the next few years. About 25 per cent of the forest is peat swamp in the low eastern plains. Only 5 per cent of the island is covered with irrigated ricefields, 8 per cent (mainly in the north-east) comprises rubber and oil palm plantations, and nearly one-third is scrub, bush, grassland or extensive upland agriculture.

Rice is the predominant food crop but cassava is important in the hilly areas and in much of the south. Logging and other timber industries are extremely important to most of the provinces, and if settlement in logged areas can be prevented (as the law requires), then the benefits of this could continue far into the future.

Sumatra has some wonderful wild areas with full complements of its large, colourful and impressive flora and fauna. There are the rarely visited western Mentawai Islands, particularly Siberut, with their unique wildlife, and the mountainous country of the main range in and around Mount Leuser National Park in the north, Mount Kerinci National Park and the Padang Highlands in the centre. Down in the south, there is Barisan Selatan National Park along the most southerly peninsula, and the fuming Krakatau volcano off the south coast. Towards the east there are a few flat wild areas such as the peat swamps of Berbak and the freshwater swamps of Way Kambas National Park, but both of these areas are under heavy pressure from potential settlers. Sumatra's wilderness is relatively easy to reach because of the island's several airports and an extensive road network.

Mount Leuser National Park

Mount Leuser National Park is one of the grandest forest areas in South-east Asia, and also one of the largest, covering nearly 9,500 square kilometres (3,670 square miles). It is horseshoe shaped and encompasses a wide range of habitats: from coastal forest along a small stretch of the west coast to the tall, non-volcanic peaks of the northern Central Range, including Mount Leuser itself, at 3,381 metres (11,092 feet) the second tallest mountain in Sumatra. The park is bisected by the beautiful Alas River, down which tourists can paddle and float on white-water rafts. In the far east, only two hours from Indonesia's third largest city, Medan, is Bohorok Rehabilitation Station where confiscated orphaned Orangutans are taken to be rehabilitated into the wild. This is adjacent to the village of Bukit Lawang and is reached easily by car or bus. The park has about 105 species of mammals and as well as rehabilitant Orangutans the visitor should see at least some of the other primates such as White-handed Gibbons, Thomas' Leaf Monkey, Siamang, Pig-tailed Macaques and Long-tailed Macaques. If time is taken to walk in the park one might be lucky to see tracks of Tiger, Elephant and Sumatran Rhinoceros, or the scratch marks made on tall trees by the Sun Bear. The bird fauna here is rich (313 species recorded) and beautiful with the drumming note of the Greater Coucal and the 'ki-au' call of the long-tailed Argus Pheasant being among the most characteristic sounds. More or less in the middle of the park is the Ketambe Research Station which has been a centre of wildlife research for twenty years.

ABOVE The white waters of the Alas are now attracting tourists who raft down with experienced guides, sometimes spotting monkeys, otters and even elephants on the banks.

OPPOSITE PAGE The Alas River runs through the park, in some places slowly, in other places swirling over rapids, and after heavy rain it flows in raging torrents. Note the brown young leaves of the common riverine tree *Pometia pinnata* on the right and the tall, open crown of a *Shorea* dipterocarp tree at the top left.

RIGHT Riverine forest showing wild bananas, rattan palms and tall forest trees.

LEFT The dapper Thomas' Leaf Monkey (*Presbytis thomasi*) is confined to the lowland forests of Aceh and northern North Sumatra.

BELOW LEFT The Siamang (*Hylobates syndactylus*) is the largest of the gibbons, and occurs in lowland forests throughout Sumatra. Adult pairs sing loud and complicated duets which are accompanied by much crashing and leaping in the canopy. They are relatively easy to approach at such times, while they are concentrating on other matters.

BELOW Orangutans (*Pongo pygmaeus*) are found in the lowland forests of the park. On the eastern side is the Bohorok Rehabilitation Station from which confiscated pet Orangutans are gradually released back into the forest.

ABOVE Sub-adult male Orangutans develop long beards and, when fully mature, flanges on their cheeks. They live solitary lives, interacting with others only when courting and mating. Even when several animals are attracted to a particular fruiting tree, there is no social interaction.

ABOVE RIGHT Young male Siamang. Note the pouch beneath the throat. This is blown up taut when he sings, and it probably acts as a resonator to increase the distance over which the song is heard.

RIGHT This Long-tailed Macaque (*Macaca fascicularis*) bares its teeth and raises its eyebrows to reveal white eyelids in a threatening pose to the photographer. This is the most common monkey in Indonesia, being found from Sumatra through to Lombok. It is generally found in riverine forests but is tolerant of secondary growth and is not uncommon near villages.

ABOVE The rare two-horned Sumatran Rhinoceros (*Dicerorhinus sumatrensis*) has one of its largest populations in Mount Leuser National Park. Despite being legally protected, it is still hunted and is now severely threatened.

LEFT Lower montane forest at about 1,800 metres (5,900 feet) showing the heavy growths of moss covering every available surface in this frequently cloudy environment.

ABOVE LEFT A species of a leguminous shrub (*Saraca*) whose bright and perfumed flowers are particularly striking in the gloom of the forest floor.

ABOVE CENTRE The dark-blue fruit of *Lasianthus* sp., a relative of coffee, is probably dispersed by birds inhabiting the understorey of the forest.

ABOVE RIGHT Male flowers of *Poikilospermum* sp., a relative of nettles and figs, which grows as a semi-strangling epiphyte and whose seeds are shot out of the fruit when it is ripe.

RIGHT *Nepenthes pectinata* pitcher plants, found in upper montane forest at about 2,000 metres (6,560 feet). The pitcher is a modified leaf-tip covered by a lid which prevents too much rain-water from entering. Insects are lured to the pitcher both by smell and the flesh-like colour, but slip on the waxy lip and fall into the fluid at the bottom where they are digested. The nutrients thus released are then absorbed by the plant.

FAR RIGHT *Aeschynanthus longiflorus* is one of many exceptionally beautiful herbs found on the forest floor or growing epiphytically on trees. This plant was found in upper montane forest at about 2,200 metres (7,220 feet). It is pollinated by small birds such as the colourful sunbirds.

RIGHT Unlike many orchids, which grow on trees, *Calanthe* orchids grow on the forest floor.

FAR RIGHT A species of *Amomum* (possibly *A. apiculatum*, a type of ginger known only from central West Sumatra).

LEFT The male Asian Fairy Bluebird (*Irena puella*) is a breathtakingly beautiful inhabitant of lowland rain forest. The female is slate-blue with dark wings. The males call with a loud and mellow note.

OPPOSITE PAGE Long, moulted feathers are all most people ever see of the shy Argus Pheasant (*Argusianus argus*). The male clears a 'dance floor' on which he displays, strutting around and lifting his tail and wing feathers into a peacock-like fan in order to impress the female.

BELOW LEFT The Malay Peacock Pheasant (*Polyplectron malacense*) is a shy ground-living bird found in Sumatra, Borneo and the Malay Peninsula.

BELOW Blyth's Hornbill (*Rhyticeros plicatus*) is one of ten hornbill species living in Sumatra, up to eight of which can be found in a single area of forest. This is the most widespread hornbill, occurring as far east as New Guinea.

The widespread fungus *Lentinus strigosus* on a fallen log starting the process of decomposition and recycling of nutrients within the forest.

Nephila maculata is an impressively large spider seen in secondary vegetation, villages and towns. As with most spiders, the male is much smaller than the female. The silk of the web is a conspicuous yellow colour.

Tortoise beetles (Cassinae, Chrysomelidae) are recognized by their shell which is formed from the enlarged forewings and modified surface of the thorax. The golden sheen of the 'shell' quickly disappears after death.

Giant harpagophorid millipedes moving determinedly across the forest floor are a not uncommon sight. These creatures feed on decaying leaves and wood, helping to recycle nutrients.

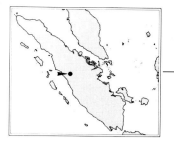

Padang Highlands

Many visitors to Sumatra take the bus from Medan, the capital of North Sumatra, south to Padang, the capital of West Sumatra, via the world's largest crater lake of Lake Toba and the Padang Highlands. This area extends from around Padang Sidempuan south of Lake Toba to Mount Kerinci, Sumatra's tallest mountain, in the south. The winding road provides some wonderful scenery with ricefields and fish ponds against a backdrop of enticing and cool-looking, misty-blue mountains. In the centre, near the hill town of Bukittinggi, known by the Dutch as Fort de Kock, are a great many sights, and it is worth lingering here for a while. At Batang Palupuh there is an accessible area of forest where the huge *Rafflesia* flower can be seen. Dramatic limestone gorges mark out the Lembah Harau Reserve to the east and also around Payakumbuh there are some caves where busy crowds of bats and swiftlets tear in and out of the blackness. Two bare-coned volcanoes, Merapi and Singgalang, stand guard on either side of Bukittinggi, and the cool and forest-fringed crater lake of Maninjau is not far away. If one continues south on the Trans-Sumatra Highway there are interesting cultural sights around Solok, and more spectacular scenery around the Sebelas Hills. That route, however, misses out the pair of lakes called Diatas and Dibawah (meaning 'Above' and 'Below' respectively), and also Lake Kerinci near Sungaipenuh, the gateway into the huge Mount Kerinci National Park which covers nearly 15,000 square kilometres (over 5,790 square miles) of lowland and montane forests.

A West Sumatran rural scene showing a typical Minangkabau house with the forest-clad Mount Merapi behind.

LEFT The Harau Valley Reserve, to the east of Bukittinggi, is famous for its beautiful limestone scenery.

BELOW The Chestnut-capped Laughing Thrush (*Garrulax mitratus*) is a popular cage bird and is trapped in its forest habitats.

LEFT View down into the crater lake of Lake Maninjau from its forested rim. A reddish-brown subspecies of the Banded Leaf Monkey (*Presbytis melalophos*) can sometimes be seen here.

RIGHT The amazing *Rafflesia micropylora* can be found in the forests of Aceh, at the northern end of the Padang Highlands.

RIGHT Many fruit of commercial importance have wild relatives in the forest. For example, this wild Jackfruit (*Artocarpus* sp.) is an important genetic resource and needs to be conserved.

FAR RIGHT *Paphiopedilum chamberlainianum* is a terrestrial orchid endemic to the highlands of Sumatra. These slipper orchids have suffered greatly at the hands of collectors but loss of their forest habitat is a still greater threat.

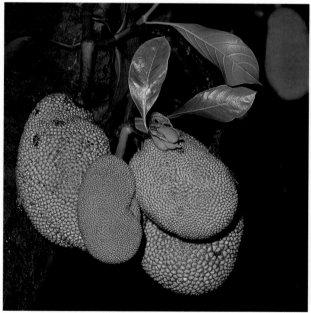

RIGHT *Phalaenopsis cornucervi* is found in lowland and hill forests from Thailand to Borneo. Although usually growing on other plants, it sometimes binds its roots around rocks on steep hillsides. Moth orchids such as this are much sought after by collectors and many wild populations have seriously diminished as a result.

FAR RIGHT The variable common Christmas Orchid (*Calanthe triplicata*) is found in scrub and forest from sea level to nearly 2,000 metres (6,560 feet), from Madagascar and India to the Pacific Islands.

LEFT The Serow or Mountain Goat (*Capricornis sumatraensis*) is one of the Sumatran mammals more or less restricted to mountains, particularly in limestone areas.

OPPOSITE PAGE Tree ferns (possibly *Cyathea contaminans*) in disturbed forest near Padang Sidempuan, West Sumatra. The trunks of these ferns are collected to make attractive containers and for substrate on which to grow orchids.

ABOVE Young Pig-tailed Macaques (*Macaca nemestrina*) are trapped in the forests, or in the fields as they make raids for food, and are then trained to pick coconuts, a job they do very efficiently.

LEFT The enormous Flying Fox (*Pteropus vampyrus*) is the world's largest bat. Colonies are found in forested areas and they can sometimes be seen in rural or even urban areas as they fly between their roosts and feeding sites. Although accused of being pests in orchards, they are very particular about eating only ripe fruit, and a conscientious orchard owner should have picked his fruit and despatched it to market before that stage.

East Coast

The great Sumatran rivers flowing into the turbid Malacca Strait bring with them large quantities of rich silt which attracts all sorts of wildlife. Much of it is not immediately visible because it is below the mud surface. Some of these inhabitants, such as peanut worms and ragworms, remain hidden except when the tides cover the mud, whereas others, such as mudskippers and crabs, emerge from their burrows as soon as they feel the coast is clear. Mudskippers can walk, run or jump on the surface of the mud using accomplished flicks of their tails. They can even climb up sloping trees using their pelvic and pectoral fins. About forty species of wading birds can be seen on the coasts of Sumatra where they feed on the small crabs, prawns, various

molluscs, larval fish and worms, probing the mud with their bills or watching for movements on the mud surface. Most of the waders are relatively small migratory species which are present between February and April and between September and November each year, as they fly to and from their breeding grounds in eastern Asia. The most practicable way to see this area is by hiring a small boat, leaving from Palembang, Jambi, Dumai or Medan.

Fishing platform in the shallow Malacca Strait offshore from Sungsang, South Sumatra.

ABOVE Grey Heron (*Ardea cinerea*) on a coastal mangrove tree.

RIGHT Milky Storks (*Mycteria cinerea*) breed in several locations along the eastern shores of Sumatra, making this area extremely important for the continued survival of the species.

LEFT The Estuarine Crocodile (*Crocodylus porosus*) has been slaughtered by man for centuries for its skin and also out of fear and hate. Large specimens do sometimes kill people.

BELOW LEFT Long-tailed Macaques (*Macaca fascicularis*) are quite common in the mangrove fringe where they eat crabs and any other small animals they can catch, as well as mangrove leaves and fruit.

BELOW Behind the mangrove trees, or upstream where the water is less salty, the stemless Nipa Palm (*Nypa fruticans*) can be found. The flowers are pollinated by small sweat bees (*Trigona*).

BOTTOM The large, globular fruit of the Nipa Palm are made up of forty to fifty seeds which break off and float in the water, germinating when they lodge against something.

Kerinci-Seblat National Park

Kerinci-Seblat extends for nearly 350 kilometres (220 miles) down the Barisan mountains covering 14,846 square kilometres (5,732 square miles) in four provinces: West and South Sumatra, Jambi and Bengkulu. In the centre of the park is a rift valley containing a major enclave with the town of Sungaipenuh, coffee plantations and Lake Kerinci which is covered by the introduced Water Hyacinth. The dominant feature is the volcanically active Mount Kerinci itself, which at 3,805 metres (12,483 feet) is the highest Indonesian mountain outside Irian Jaya. To the east is 'Lake' Bento, the highest freshwater swamp in Sumatra and very interesting botanically, even if it has been disturbed. Behind this is Mount Tujuh with a beautiful and almost untouched crater lake below it. As with many of the other conservation areas in Indonesia, the reason it is still forested is because it is extremely steep and unsuitable (if not impossible) for farming and other development. The park contains some little-known lowland dipterocarp forest with many species of *Shorea* and *Dipterocarpus*, as well as the giants of the plant kingdom, *Rafflesia arnoldi* and *Amorphophallus titatum*. It also boasts all the large Sumatran animals, except Orangutans: Elephants, Rhinoceros, Tapirs, Tigers, Sun Bears and Clouded Leopards. It is hoped that conservation officers will soon be installed to assist the local park staff to increase awareness and improve management of this exceptional area.

The remote crater lake on the side of Mount Tujuh, east of Mount Kerinci, is at 1,996 metres (6,548 feet) and has an area of 10 square kilometres (4 square miles). It is the least disturbed major lake in Sumatra, and so far has no introduced fishes or water plants.

LEFT View to the smoking Mount Kerinci, at 3,805 metres (12,483 feet) the highest mountain in Indonesia outside Irian Jaya.

BELOW *Plectranthus javanicus* is a distant relative of mint and thyme, but smells foetid rather than aromatic when bruised. It is found in forest edges, along streams and in clearings.

ABOVE *Rhododendron retusum* grows on mountains between 1,500–3,400 metres (4,920–11,155 feet). It is pollinated by small sunbirds (*Aethopyga*), and is one of twenty-six species of *Rhododendron* known from Sumatra.

LEFT Lowland rain forest in Kerinci-Seblat National Park.

RIGHT The summit slopes of Mount Kerinci are devoid of vegetation, partly because of the sulphurous air and partly because of the frequent falls of fresh, hot ash.

BELOW *Phaius flavus* is a large terrestrial orchid found in mountain forests from India to New Guinea and Taiwan.

ABOVE *Polygala venenosa* is a small shrub of the understorey. The appendage to the lowermost petal changes colour with age from yellow to red.

RIGHT Timbulan Waterfall can be seen near the road between Sungaipenuh and Padang not far from Kerinci-Seblat National Park.

Way Kambas National Park

Way Kambas lies at the south-east tip of Sumatra on the low coastal plain. It has been a conservation area for nearly seventy years but has suffered terribly from the pressures of Java's population expanding to the west. It covers 1,300 square kilometres (500 square miles) but only about 20 per cent of this is still forest and virtually all of it has been logged to some extent. Nevertheless, it contains the largest area of freshwater (i.e. non-peat) swamp forest remaining in Sumatra, it has a wide range of habitats, and its closeness to Java which has caused it so many problems also increases its importance because of its potential for recreation and education. It is important that protection is afforded to the remaining wild areas within the park so that they will regrow and again support a rich diversity of animals and plants. Although the Javan Rhinoceros was lost from here many decades ago, there are still small populations of

Tiger, Sun Bear, Wild Dog, a number of civets, deer, Tapir, various primates, and interesting birds such as the White-winged Wood Duck and Milky Stork. In addition, the en-dangered thin-snouted crocodile known as the False Gharial is also found here, albeit in small numbers. The largest mammal is the Elephant which is particularly common because of drives organized by the local government to force these animals out of neighbouring agricultural areas. Some elephants are kept at a training centre, where they learn to carry visitors and to do simple tasks in the hope that people will see them as useful and valuable beasts rather than undesirable pests.

View of the park from the Way Kanan River. Characteristic of these eastern swamplands are the tall Serdang palms (*Livistona hasseltii*).

BELOW One of the trained elephants (*Elephas maximus*) in the park bathing at the end of the day.

RIGHT View of a disturbed part of the park. Way Kambas has suffered greatly from illegal logging and other wood collecting, but there is every reason to suppose that protection of the remaining trees would allow the forest to recover, given time.

BOTTOM It is hoped that the group of elephants trained at the park will find gainful employment, perhaps conveying tourists, thereby demonstrating their usefulness.

RIGHT The Tapir (*Tapirus indicus*) is found only in southern Sumatra, and primarily in low-lying or swampy forests. This species is not intolerant of logging, and can even benefit to a certain extent because its major food plants are those succulent species that grow up after the forest has been disturbed.

BELOW *Kadsura scandens* is a primitive climbing shrub, related to the magnolias, whose flowers are pollinated by small beetles. Its stems are a source of pure drinking water, used by forest travellers.

BOTTOM The bracket fungus *Microporus xanthopus*, a widespread tropical species.

LEFT *Cookeina tricholoma*, a cup fungus growing on the spiny decomposing trunk of the Nibung Palm (*Oncosperma tigillarium*).

BELOW Stick insects are among the most remarkable of the jungle's oddities, some growing up to 30 centimetres (12 inches) long.

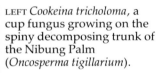

LEFT A Sambar hind (*Cervus unicolor*) in riverine forest. Though this large animal is protected by law it is a popular sporting target. The excursions of nocturnal hunting parties in off-road vehicles with powerful spotting lamps cannot always be prevented by the poorly resourced forest guards.

BELOW A female Common Goldenback Woodpecker (*Dinopium javanense*), a species found in open forests, plantations and gardens. The male is similar but has a red crest.

BOTTOM Night Herons (*Nycticorax nycticorax*) are probably as abundant as other more conspicuous egrets, but because of their largely nocturnal habits are less noticed. Around dusk, flocks can be seen flying slowly from their roosts to paddy fields and marshes to feed on frogs, small fish and insects.

BELOW The Purple Moorhen (*Porphyrio porphyrio*) may seem a heavy and cumbersome bird, but it can walk around easily on lily pads and in reed beds where it grasps the tall stems with its long toes, flicking its tail the while to reveal the white feathers beneath. It rarely flies, preferring to skulk around in swampy vegetation.

Krakatau

Krakatau is one of the most famous volcanoes in the world. In 1883 it erupted with a force equivalent to 2,000 Hiroshima bombs; the thunderous roar was heard as far away as Sri Lanka, the Philippines and Australia. Ash and pumice shot 80 kilometres (50 miles) into the sky and in such quantities that the skies were dark during the day and people in Jakarta and Bandung had to use lamps. Enormous tidal waves were created by the disturbance, detected by instruments in Alaska and South Africa, and 36,000 people lost their lives. Krakatau is in fact the name given to a group of four islands representing the remains of a single prehistoric cone split asunder by earlier eruptions. The islands have changed their shapes over time, and the newest island, Anak Krakatau, emerged from the sea only in 1930. The ecological interest of Krakatau comes from the fact that the land was totally sterilized during the 1883 eruption and it is possible to study the patterns by which life has colonized the virgin land. All the plants and animals on the islands have arrived by floating or in the air. The species present were first surveyed in 1886, and expeditions have continued to visit the islands since then to observe changes. Since about 1950 successive expeditions have found some 200 plant species, although the actual species do not always remain the same; new ones arrive, others fail to establish themselves. The plants found there are weedy species typical of open land and as yet only one species of forest tree has been found; it will take centuries or millennia before forest is re-established. Krakatau is part of the Ujung Kulon National Park and can be visited on tours leaving from Labuan on the adjacent West Java coast.

OPPOSITE PAGE View from Anak Krakatau, which first emerged from below the sea in 1930, showing in the distance the She-oak trees (*Casuarina equisetifolia*) which were among the first species to colonize the bare strand.

RIGHT Apart from the coastal fringe, Anak Krakatau is more or less devoid of vegetation.

BELOW Some of the lava slopes on Anak Krakatau have begun to be colonized by hardy grasses and ferns.

Focus on Kalimantan

Kalimantan is that major part of the large island of Borneo that is under Indonesian jurisdiction, south of the Malaysian states of Sarawak and Sabah and the small but very rich country of Brunei Darulsalam. Borneo is the third largest island in the world and, with the exception of the swampy southern areas, much of the land is either hilly or mountainous. The mountains in Kalimantan, primarily in the centre and north, are considerably smaller than those in the other large Indonesian islands, with the highest ones being apparently unnamed peaks rising to over 2,500 metres (8,200 feet) on the border between East Kalimantan and Sarawak. Extensive lowland plains do exist, notably around Lake Sentarum just below Sarawak in the west, and Lake Semayang in the east. The longest rivers in Indonesia are in Kalimantan with the westward-draining Kapuas over 600 kilometres (375 miles) long, and the eastward-draining Mahakam and the southward-draining Barito, some 500 kilometres (300 miles) long. A classic fan-shaped delta is found at the mouth of the Mahakam.

Kalimantan has a total area of 535,834 square kilometres (206,886 square miles), about one-third of the area of the entire country. It is divided into four provinces, three of them larger than Java and Bali combined: East Kalimantan (202,404 square kilometres or 78,148 square miles), the country's second largest province after Irian Jaya and tremendously rich in timber, gold, oil and gas; West Kalimantan (146,760 square kilometres or 56,664 square miles) much deforested but still providing quantities of timber; Central Kalimantan (152,600 square kilometres or 58,919 square miles) with extensive mangrove and peat swamps and heath forests, much of which is incapable of supporting sustainable agriculture; and the relatively diminutive South Kalimantan (37,660 square kilometres or 14,540 square miles) with most of its forests felled and replaced by erosion-prone, hot and unproductive grasslands.

The population of Kalimantan is 8.9 million, growing slightly faster than the national average because of the effects of the government transmigration programme and the spontaneous immigration of people from crowded Java and Bali who seek land on which to start new lives. The indigenous people are often erroneously lumped together under the term 'Dayak' but they actually comprise a mosaic of groups including the Iban, Kayan, Kenyah, Ngadu and Land Dayak. Most of these peoples once lived in longhouses along the large rivers from which they would hunt, fish and collect food and other forest produce (such as beeswax, hornbill ivory, rattan cane, resin and camphor). Any surplus not needed for daily living would be traded for other goods, such as beads, ceramic jars and salt, at the coast and other trading stations. Some old longhouses measured up to a kilometre (nearly two-thirds of a mile) in length, made up of separate rooms for each nuclear family; those found today are rather more modest. The wide verandahs provide communal living space for chatting, weaving, playing and singing. After a couple of decades of discouraging the use of these traditional homes, the government now has a much more relaxed attitude and longhouses are being built again. The widespread use of outboard motors and the ubiquitous *klotoks* or motorized canoes have made communication between settlements much easier than in the past, but children in the remoter areas have to stay away from home for long periods to attend high school. Some tribal groups remain. One of these, the Punan, used to wander from one temporary settlement to the next, hunting and gathering food, and kept out of the way of the other groups. Today, almost all the Indonesian Punan have settled and their traditional nomadic way of life is followed by very few.

Most of Kalimantan has a hot and wet climate all year round. There are two wet seasons, the major one falling between November and April. The mean annual rainfall is about 3,000 millimetres (120 inches), and most places receive between 2,000 and 4,000 millimetres (80 and 160 inches). The driest and most seasonal parts of Kalimantan are along the east coast, but the area is served by the huge Mahakam River and its tributaries, so the

The Kahayan River near Palangkaraya, the capital of Central Kalimantan.

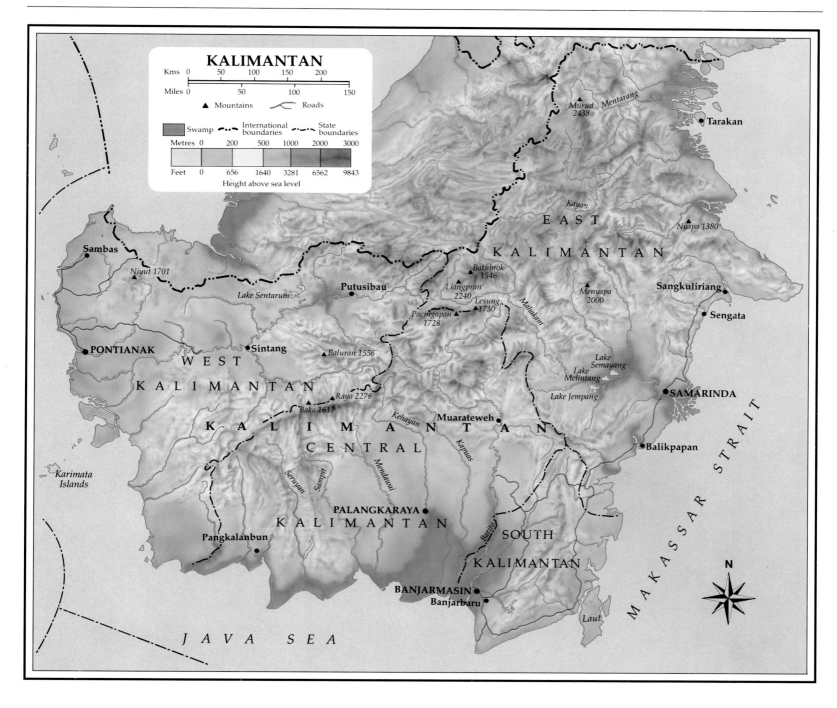

KALIMANTAN

Kms 0   50   100   150   200

Miles 0   50   100   150

▲ Mountains   Roads

Swamp   International boundaries   State boundaries

Metres 0   200   500   1000   2000   3000

Feet 0   656   1640   3281   6562   9843

Height above sea level

forest is very similar to that found in the lowlands over the rest of the island.

Just under three-quarters of Kalimantan still has some form of forest cover, but much of this has already been logged at least once. About 10 per cent of the land comprises the peat swamp forests of Central Kalimantan. Over one-fifth of the land is under scrub, bush or different forms of shifting cultivation, and only about 4 per cent is under irrigated rice and tree crops. Rice is the predominant food crop but much of this comes from the hill rice rather than paddy fields. Cassava is also common, as is taro in the hinterlands. Forest industries are more important here than anywhere else in the country, but there are few signs yet that this can be sustained in the long term. Forest plantations have begun to be developed and Central and South Kalimantan have important rattan plantations. Minerals such as diamonds, gold and coal, as well as oil and gas, are also important. East Kalimantan produces nearly half of the country's gas and one-sixth of its oil.

There are still vast areas of wild country in Kalimantan, particularly inland, though much of it is somewhat difficult to reach since the road network is skeletal and there are few commercial airports. River boats provide the major means of transport. One of the largest wild areas in Indonesia is the Kayan-Mentarang Reserve bordering Sarawak in northern East Kalimantan which has an area of some 16,000 square kilometres (6,180 square miles). The forests around Mount Bentuang and Karimun in western East Kalimantan, also on the Sarawak border, are extremely interesting botanically and zoologically. Much of these areas is hill forest, and lowland forest is perhaps best represented in the Bukit Raya/Bukit Baka area in Central and West Kalimantan, and Mount Palung in western West Kalimantan. All the above areas are relatively inaccessible, and so many visitors go to Tanjung Puting National Park in the south, which has a mosaic of lowland habitats including swamps and heath forests, and to Kutai National Park north of Samarinda in East Kalimantan.

Tanjung Puting National Park

Tanjung Puting National Park (3,050 square kilometres or 1,178 square miles) lies on the south coast of Kalimantan and constitutes an area of seasonally inundated low-lying land of peat swamps and heath forests. The heath forest found in the north of the park is interesting because it grows on poor white-sand soils and characteristically has only medium-sized trees, most of them with rather small leaves and a few orchids or ferns growing on their boughs. In addition, there are generally many insectivorous plants, a few climbing plants, and a rather low density of animal life. The peat swamp forest is in the middle of the park and many of the trees have stilt roots and aerial roots as adaptations to the frequent flooding. The best known animals are the Orangutans, made famous by the efforts of Dr Birute Galdikas over the last twenty years to rehabilitate confiscated specimens at the Camp Leakey Research Station at the south of the park. The next largest primate present is the bizarre Proboscis Monkey. Six other primates can be found, as well as Clouded Leopard, deer, civets and Mouse Deer. There are over 200 species of birds known from the park including many wetland species and a good range of forest birds. In the water, threatened species include the Asian Bonytongue fish and the False Gharial crocodile. The park has suffered considerable disturbance and over a third of it has been reduced to scrubland by logging and other activities.

Forest at the edge of the Sekunir River near to Camp Leakey. The forest fringe is dominated by Screwpines (*Pandanus*) and by the long blade leaves of *Hanguana malayana*.

ABOVE The tranquil blackwaters of the Sekunir River.

ABOVE Purple Heron (*Ardea purpurea*) watching the river below for potential prey.

BELOW Monitor Lizards (*Varanus salvator*) are often seen by river banks but they are shy of people and tend to race into thick vegetation as soon as they are disturbed.

BELOW Sun Bears (*Helarctos malayanus*) are among the most feared of the forest animals. They are short-sighted and if surprised may be aggressive, rising up on their hind legs and slashing out with their long, sharp claws. Even cubs like this can be dangerous.

OPPOSITE PAGE, TOP LEFT The Agile or Black-handed Gibbon (*Hylobates agilis*) is found in south-west Kalimantan. In the headwaters of the Mahakam, Kapuas and Barito Rivers there is a zone of overlap between this species and the Bornean Gibbon (*H. muelleri*), and hybrid animals occur.

OPPOSITE PAGE, TOP RIGHT Tanjung Puting National Park is best known for its Rehabilitation Centre where Orangutans confiscated by the government are trained to fend for themselves in the forest. Its headquarters, Camp Leakey, also serves as a research centre for Indonesian and foreign students. BOTTOM LEFT AND RIGHT Wild adult male Orangutans with large cheek flanges sometimes visit the camp and mate with rehabilitant females.

LEFT Part of a group of Proboscis Monkeys (*Nasalis larvatus*) preparing for sleep in their night tree by the Sekunir River.

BELOW LEFT This large species of monkey is most commonly found along coastal rivers where it feeds on leaves.

BELOW RIGHT Despite its size – males can weigh up to 24 kilograms (54 pounds) – the Proboscis Monkey is adept at leaping between tree canopies.

LEFT Wild Bilberry (*Vaccinium* sp.) on the bank of the Sekunir River. From a distance, the flush of new leaves resembles pink flowers.

BELOW LEFT *Vanda hookeriana* is a flamboyant wild orchid found in sunny positions in swamps, often with the sword-leaved *Hanguana malayana*, in Peninsular Malaysia, Sumatra and Borneo.

BELOW The seeds of a *Gardenia* are attractive as food to birds in the forest understorey.

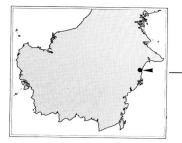

Kutai National Park

Kutai National Park in East Kalimantan has had a troubled history. Originally gazetted fifty-five years ago as a 3,000-square kilometre (1,160-square mile) game reserve, it was reduced by a third in 1971 to allow for logging. Further logging and land clearance for small-scale agriculture damaged the remaining reserve and in 1982/83, just after it was declared a national park, massive fires swept through the area burning 60 per cent of the primary and secondary forest after a severe drought exacerbated by the various forms of disturbance. As if that were not enough, a fertilizer factory and a coal mine have been developed within the park boundaries, and a major road through the park is likely to have serious impacts. Yet, despite all this abuse, Kutai still has a great deal to offer for conservation and research. Eleven primate species have lived in the park but it is not yet known for sure how hard they have been hit by the fires and drought.

Other notable species include the Banteng, deer, Sun Bear and a variety of smaller carnivores. Again, before the worst of the damage, about 300 species of birds were known from Kutai, including eight species of hornbill. The priority now is to protect all the remaining forest areas as important sources of seeds from which new trees will grow in the damaged areas, the regeneration of which also needs to be carefully nurtured. Illegal settlers (most of them recent migrants from neighbouring Sulawesi) within the park will probably have to be translocated to other areas.

The forests along the Sangata River have been logged but are still of some value to wildlife.

LEFT A female Orangutan smelling the flowers of a climbing vine. These animals have a very catholic diet, eating all sorts of leaves, fruit, flowers, bark and small animals.

BELOW Five-bar Swallowtail (*Graphium antiphates itamputi*).

BOTTOM LEFT The Rajah Brooke's Birdwing Butterfly (*Trogonoptera brookiana*) is one of the most exciting sights in the forest. As it flies by, the green of its 17-centimetre (7-inch) wings appears intensely bright, contrasting markedly with the purple-red saddle on its thorax. Generally only the males are seen, in forest clearings and along logging roads, the females frequenting the high canopy.

BOTTOM RIGHT A female Malay Baron (*Euthalia monina bipunctata*), one of the most unusual butterflies in the world. The females are always the same but the males have a wide variety of colour forms. In all other butterflies with more than one colour form, it is the female which varies.

TOP LEFT and RIGHT Bush crickets or katydids (Tettigonidae) are recognized by their long jumping legs and antennae. The female on the left bears a sickle-shaped ovipositor which is used for cutting slits in plant stems before laying her eggs.

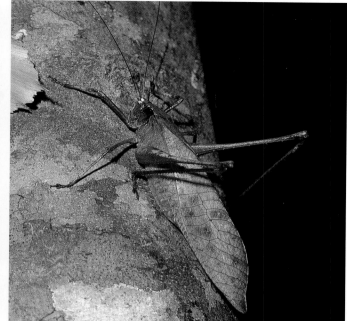

CENTRE LEFT A cricket (Gyrillidae) about to take off from its leaf perch.

CENTRE RIGHT A nymph of a forest hemipteran bug.

BOTTOM LEFT A group of heteropteran shield bugs (Catacanthus incarnatus).

BOTTOM RIGHT The bright colours of hemipteran bugs such as this warn potential predators that they should be left alone. If molested they can exude a most unpleasant smell.

ABOVE Gaudy, predatory assassin bug (Reduviidae) which feeds on other insects.

ABOVE RIGHT The Jewel Beetle (*Chrysochroa fulminans*) is a sun-loving insect which feeds on leaves.

ABOVE A flat-backed *Platyrhachus* millipede found on the forest floor under the bark of rotting trees where it feeds on fungi rather than the wood itself.

Lantern Bugs (Fulgoridae) are little-known relatives of the cicadas. The function of the peculiar projection above the head has not yet been explained.

Trilobite larva of a lycid beetle (*Duliticola*), a relative of the firefly beetles. Very little is known about these creatures, except that they exude digestive juices onto the decaying wood on which they are found and then suck up the resulting 'soup'.

ABOVE The Crested Lizard (*Calotes cristatellus*) is quite common around the forest edge and even in gardens. It is able to change its colour rapidly, and is an accomplished climber, using its long tail to balance.

RIGHT Monitor Lizards (*Varanus salvator*) can grow to nearly 2.5 metres (8 feet) and are tenacious predators.

ABOVE The Rhinoceros Hornbill (*Buceros rhinoceros*) is one of eight species of hornbill found in Kalimantan. It is the most important species culturally and its carvings and image can be seen on longhouses and public buildings.

RIGHT Sambar Deer (*Cervus unicolor*) are more common than most travellers might imagine because they are generally nocturnal and are very shy, a result of centuries of hunting.

Fungi are of major importance in the decomposition and recycling of organic material in the forest: ABOVE *Amanita aporema*, TOP RIGHT *Mycena* sp. and BOTTOM RIGHT *Marasmius congregatus*.

LEFT Despite its ginger-like appearance, *Phrynium* belongs to another family (Marantaceae).

BELOW LEFT Gingers are a common and diverse group found on the forest floor and at the forest fringe. The red flowers belong to an as-yet unnamed species of *Etlingera* first collected in Sabah, where *Amomum laxisquamosum* (BELOW) is also found.

ABOVE The different life-forms in a forest never fail to fascinate, as here in the new shoot on a *Leea* shrub.

ABOVE RIGHT This *Dendrocnide* is commonly felt before being seen. It is a relative of the stinging nettles, and the hairs on its leaves and stems deliver an extremely painful sting that may result in swelling of the lymph glands.

RIGHT This wild species of arrowroot, *Tacca integrifolia*, is a splendidly bizarre lily-like plant found on the forest floor.

Focus on Java and Bali

Volcanoes dominate the landscapes of both Java and Bali. On Java they run like a spine down the back of the island, and on Bali form a focus in the north towards which the rest of the island looks. Java is the most volcanically active island in the whole of South-east Asia: it has twenty-three of the seventy-eight Indonesian volcanoes classified as active (those that have been active since 1600), and half of its land area comprises recent volcanic deposits. The remainder is up-raised marine deposits or alluvial plains. As a result of these volcanic deposits, both Java and Bali are exceptionally fertile and productive and therefore intensively farmed. Java has a total area of 138,204 square kilometres (53,360 square miles) divided into two special areas and three provinces: the Special Capital Area of Jakarta (590 square kilometres or 228 square miles); the Special Area of Yogyakarta (3,169 square kilometres or 1,223 square miles); West Java (46,300 square kilometres or 17,876 square miles); Central Java (34,206 square kilometres or 13,207 square miles); and East Java including Madura Island (47,922 square kilometres or 18,503 square miles). Bali (5,561 square kilometres or 2,147 square miles) is the smallest province in the country.

Java is one of the most densely populated places in the world, with some 109 million inhabitants. There are only two ethnic groups on Java, the Sundanese and Javanese, but there are a number of cultural areas. Thus there are Sundanese people in western Java, Madurese on Madura, and three groups of Javanese: one along the central and eastern north coast, one in the central heartland, and one to the east in the historic area of Blambangan. Among the Sundanese there is a group of highly isolationist and secretive people, known as the Badui, in the western highlands of West Java, who have very strict moral and farming codes; and among the Blambangan people there is an upland group, the Tenggerese, around Bromo volcano, with a relatively pure old Javanese, pre-Islamic culture.

The earliest human remains in Asia were found in Central Java last century and this 'Java Man', *Homo erectus*, is estimated to date from about one million years ago. Since then Java has been subject to waves of migration from mainland Asia, and within the last millennium the great kingdoms adopted Buddhist, Hindu and Islamic religions as well as incorporating elements of local beliefs. It is generally believed that Javanese Hindu noblemen fled to Bali some five hundred years ago to escape the spread of Islam and enriched the Hindu religion which persists there to this day among almost all its 2.6 million inhabitants. Aesthetically pleasing temples, often in dramatic settings, and colourful votive offerings carried on the head by classically adorned women are typical sights sought by the large number of visitors to this island.

Java is by far the most industrialized and developed island in the archipelago, and these factors, together with the fertility of the land and the very high population densities, combine to make the environmental problems the most acute in the country, from the point of view of pollution, floods, droughts and of forest loss. Most of the industries are sited along the north coast, but there is also a major centre at Cilacap at the west end of the south coast of Central Java. Gold is mined in West Java and oil is drilled just off the north shore, but such contributions are small compared with the production from Sumatra and Kalimantan. Madura, the large limestone island to the north-east of Java, has no forest left, and Java retains only about 10 per cent of its natural forest. This has resulted in the inevitable loss of animal and plant species and of a wealth of potentially useful genetic resources. The forests that do remain are mainly on the mountains, and act as refuges for many of the endemic species. Nearly a quarter of the island is devoted to wet ricefields, 13 per cent to settlements, and 18 per cent to tree crops (primarily teak), the largest percentages in the country.

There are perhaps more strikingly different climate types on Java than on any other island in Indonesia. In the west and the central mountains, annual rainfall can reach a staggering 7,000 millimetres (280 inches) and Bogor, the town nestled between Salak and Gede-Pangrango volcanoes just south of Jakarta, achieves fame in the *Guinness Book of Records* as having the most thunderstorm days in a year. Although wet and dry seasons do exist, they are not always particularly well defined and can vary quite considerably in their timing from year to year. Towards the east, however, the climate becomes very dry and seasonal, with the driest time being from July to September. Here barely more than 1,000 millimetres (40 inches) of rain falls a year, and the countryside has a very different look from that in the west. Bali is drier on average than Java with an average annual rainfall of 1,900 millimetres (75 inches) against 2,576 millimetres (100 inches) for Java, but is wetter and less

The many volcanoes on Java and Bali have produced very fertile soil from which, with the hard work of the farmers, bounteous rice crops are harvested. This scene is on the edge of Mount Halimun Reserve in West Java.

seasonal than neighbouring East Java.

Despite the industry and high population density, there most certainly is some wild Java, albeit rather restricted in area. In the west there is lowland forest in the Ujung Kulon National Park, a peninsula that took the full force of the enormous Krakatau explosion and which now is the sole home in Indonesia for the Javan Rhino. In the east, there are patches of lowland rain forest along the south coast, notably in Alas Purwo National Park and Meru Betiri National Park, where the Javan Tiger has almost certainly become extinct but much else still remains. The small island of Rambut, not far from Jakarta, is a major breeding and roosting site for a host of water birds. The Halimun Mountains due south in West Java get their name from the grey mist that seems to perpetually enshroud them, and local beliefs in the spirits within them have saved them from being cleared. Further east in West Java are the tall twin, forested volcanoes of Gede-Pangrango which have considerable biological interest. Mount Bromo in the east is justly famous for the sheer spectacle of the sunrise from its crater rim. Finally, there is the interest of the savannah landscape of the dry Baluran National Park in the far north-east corner where grazing animals and peacocks can be seen with ease. Bali has proportionately more forest than any province in Java, some 30 per cent of the land area. There are two major wild areas, Bali Barat National Park in the west and the forests in the centre of the island around Mount Batukau and the three lakes of Tamblingan, Buyan and Bratan. Neither of these areas is visited to any extent but they do offer great landscapes and wildlife interest.

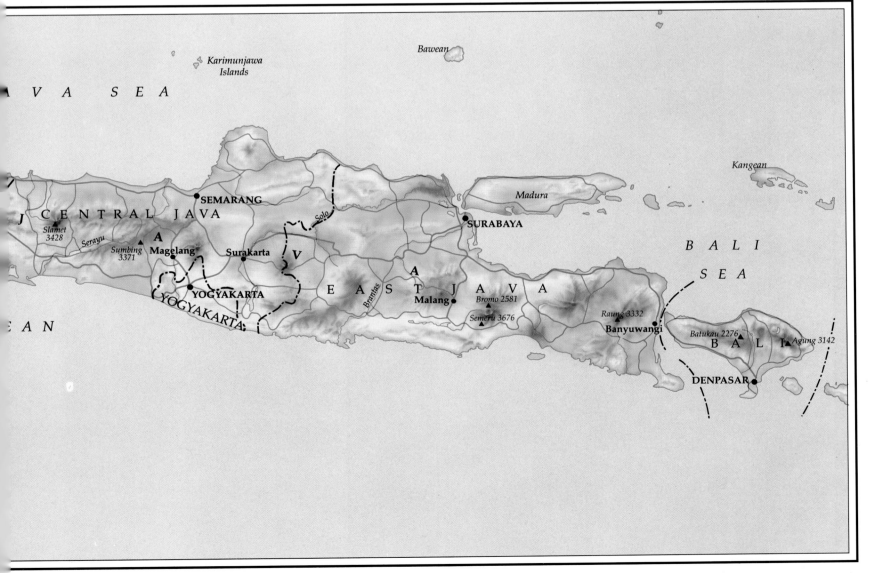

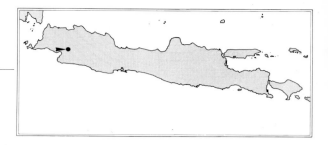

Mount Gede-Pangrango National Park

Mount Gede-Pangrango is one of South-east Asia's classic reserves where a great deal of important biological and conservation research has been conducted over the last century. It is 150 square kilometres (nearly 60 square miles) in area and is centred on two volcanoes, Mount Gede with its gaping crater and the slightly taller Mount Pangrango. To the north of Mount Gede is a field known as the *alun-alun* where vast numbers of the Javanese Edelweiss grow in a situation not unlike the mountains of the temperate zones. Small altars and recent votive offerings may be seen from worshippers faithful to ancient Hindu-influenced beliefs. Lower and upper montane as well as subalpine forests are represented here and have been well studied. The park has an outstanding list of species known only from within its boundaries, but to a certain extent this is an artefact of the disproportionate time people have spent here looking at its wildlife, be it moss, toads, land crabs, butterflies, or trees. The area is very accessible, being only just over two hours' drive from Jakarta. The best entry point is from the mountain botanic gardens of Cibodas where the visitor may become acquainted with the labelled trees before venturing up the mountain trails. The four-hour walk to the summits is generally begun before dawn so that a view is gained before the lower slopes are obscured by clouds.

The fuming crater of Mount Gede.

ABOVE Early morning mist in the moss-strewn montane forest near the hot springs on Mount Pangrango.

RIGHT The *alun-alun* field at about 3,000 metres (9,840 feet) is scattered with many light-green clumps of Javanese Edelweiss (*Anaphalis javanica*).

BELOW RIGHT The Javanese Edelweiss (*Anaphalis javanica*) is one of the most characteristic plants found at the tops of volcanoes in Java, South Sumatra, South Sulawesi and Lombok. The flowers are generally seen between April and August and are visited by many bees, flies and butterflies on sunny days. Although most specimens are less than a metre (3 feet) tall, this giant daisy is capable of growing up to 8 metres (26 feet) tall.

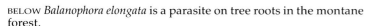

BELOW *Balanophora elongata* is a parasite on tree roots in the montane forest.

Mount Halimun

Mount Halimun Reserve is unjustly regarded as a poor cousin in Java's system of conservation areas. Despite its proximity to the very well-studied Gede-Pangrango National Park and to Bogor, the nation's premier centre of biological research, very little is known about it. Its mist-enshrouded mountain tops, reaching to 1,929 metres (6,329 feet), and its deep valleys both symbolically and in reality hide a great deal that remains to be discovered. Yet about half of its 400 square kilometres (155 square miles) comprise probably the best and most complete forest area in Java. Its lower zones hold apparently secure populations of the Javan Gibbon, Javan Leaf Monkey and other endemic species, and about half of the 145 bird species known from the reserve are rarely seen elsewhere in Java. It also protects the watershed for millions of people living to its north. At present, only three wardens are assigned to the reserve, and it is recognized that it

is time to afford the reserve the attention and protection it deserves. This reserve cannot retain its importance if lack of alternatives forces wood to be poached, birds to be trapped for sale to 'nature lovers' in the markets of the large towns, and the gold mine in the west to expand its operations towards the centre of the reserve. The indigenous people living around the area are very conservative and there are many prohibitions relating to the planting, harvesting and sale of rice, but prohibitions relating to conservation ethic have yet to be developed.

View across a clearing caused by a natural tree-fall in Halimun Reserve.

ABOVE The Bird's Nest Fern (*Asplenium nidus*) is a large and widespread epiphyte on forest trees. Dead leaves and other litter from the canopy fall into the centre of the plant and through this grows a new rosette of leaves which traps the litter above the older leaves. The roots of the fern then grow into this mat to obtain nutrients.

ABOVE RIGHT The Resam Fern (*Gleichenia*) is common in clearings, areas abandoned after cultivation, and at the forest edge. Fossils of similar plants have been found in Britain and Greenland from about 100 million years ago.

RIGHT A forested valley in the reserve with a tree fern in the foreground.

LEFT Busy Lizzie (*Impatiens* sp.) grows as a wild plant in both forested and disturbed environments. This ubiquitous plant occasionally grows epiphytically, as seen here, under moist conditions.

BELOW LEFT *Vanda tricolor* is one of Java's most showy orchids and is often cultivated as an ornamental plant. The attractive flowers also have a pleasant perfume.

BELOW CENTRE Wild bananas (*Musa* sp.) in front of the common weed *Eupatorium odoratum*.

BELOW Epiphytic *Aeschynanthus*, the flowers of which are pollinated by small birds.

LEFT The Short-tailed Green Magpie (*Cissa thalassina*) is a beautiful bird of montane forests. Despite its bright plumage, it is surprisingly difficult to see, as it skulks around in thick vegetation in the middle and upper storeys of the forest, making only short flights.

BELOW The Javan Gibbon (*Hylobates moloch*) is now restricted to a few areas in West and Central Java. Mount Halimun is its most secure site, but as it is not found above about 1,200 metres (3,937 feet) its habitat is a mere thin ring around this mountainous reserve.

Ujung Kulon National Park

Ujung Kulon National Park comprises the rectangular peninsula at the western tip of Java along with the adjoining Mount Honje region on the east of the isthmus. Ujung Kulon is possibly Indonesia's best known national park because of its famous inhabitant, the Javan Rhinoceros, which was thought, until recently, to survive only here. There is now known to be a small second population in Vietnam. The elusive rhinoceros apart, the 786 square kilometres (303 square miles) of the park have much else to offer, such as the herds of cattle known as Banteng, the Peafowl, the deer, Leopard, Wild Dog, Javan Gibbon and three other primates, excellent turtle nesting beaches, and also some wonderful scenery first noted 150 years ago by the pioneer naturalist, F.W. Junghuhn. The area was made a nature reserve in 1921 and comprises mangroves, coastal forests and various types of lowland forest. The forest on the flat areas is largely of the mature secondary type and there are a number of tree species which are known only from the park. There are also grasslands which are used for grazing by the Banteng and deer. Access is hardly simple, requiring the hire of a boat from Labuan on the west coast of Java, but the park is becoming increasingly popular and tourist facilities are steadily improving. The World Wide Fund for Nature has assisted the park in many ways over the past two decades, and an education officer has recently been assigned to work with local staff.

The beautiful rocky coast of Ujung Kulon National Park.

ABOVE Sambar Deer (*Cervus unicolor*) come down to the beach to drink, not for the water but for the salt. All herbivores need salts and these are obtained either from mineral-rich rocks, mineral water springs, or from the sea.

BELOW The elusive Javan Rhinoceros (*Rhinoceros sondaicus*) has its major population in the park. About fifty rhino live here but they are extremely difficult to see in the thick undergrowth.

BELOW A handsome Barking Deer or Muntjac stag (*Muntiacus muntjak*). This species would once have been an important prey for the now extinct Javan Tiger, and is still taken by the Leopard, the largest predator in the park.

ABOVE There are a number of locations within the park where Banteng (*Bos javanicus*) can be seen. The cows and calves are brown, the bull is black, and both have the distinctive white rump.

BELOW A widespread wild ginger *Etlingera littoralis*.

BELOW A terrestrial crab. This enigmatic specimen is possibly a *Neosarmatium*, but is unlike any known species.

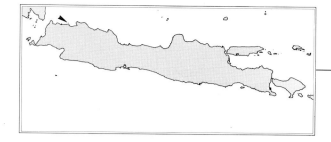

Rambut Island

Rambut Island is one of the southernmost islands in the Seribu archipelago that lies just north of Jakarta. It is a forested coral atoll and, although only 25 hectares (about 60 acres) in area, it is an important breeding site for water birds and has been a nature reserve since 1937. The northern part of the island is dominated by tidal mangrove forest, but the southern half is dryland forest, only occasionally inundated, and dominated by *buta-buta* trees. The fringe of vegetation seen as one approaches the island is of She-oak or Casuarina. An observation tower near the middle of the island provides good views of the birds flying around and overlooks the main breeding colony of herons, Oriental Darters and Great Egrets. It also provides good views of the 2,000 to 3,000 Flying Foxes as they head to the mainland to find food. The most abundant birds are the Black-crowned Night Heron with about 4,000 individuals, Little Black Cormorant with about 3,000, Purple Heron with 800, and various white egrets with a total of nearly 3,000. Perhaps the most exciting bird to visit the island is the Milky Stork which was first recorded breeding here in 1974; there are now about ten to thirty pairs on Rambut, which is probably the only breeding site for this species left on Java. The island is well protected and the threats facing the birds are on the mainland rather than here: the wetlands either side of the road leading to Jakarta's Soekarno-Hatta International Airport, which are major feeding grounds, are rapidly being converted to industrial estates and the result almost inevitably will be a decrease in Rambut's bird populations.

One of the most common tree species is the locally named *buta-buta* (*Excoecaria agallocha*). The leafless trees may have been attacked by the caterpillars of a noctuid moth which occasionally undergoes a population explosion.

ABOVE Great Egrets (*Egretta alba*) are one of the major breeding species on the island.

BELOW Purple Herons (*Ardea purpurea*) at the heronry. Several hundreds of these birds nest on the island each year, though their numbers fluctuate considerably in relation to the prevailing and previous availablity of food.

BELOW Large fruit bats (*Pteropus vampyrus*) roost in great numbers, leaving the island each dusk to find food on the mainland.

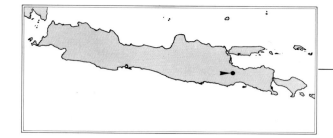
Bromo-Tengger-Semeru National Park

Bromo-Tengger-Semeru National Park is one of the more accessible of the series of great volcanic areas in Java. The Bromo area has attracted the attention of tourists because of the dramatic 10-kilometre (6-mile) wide Tengger Caldera within which lies the Bromo Crater. Visitors travel by jeep or horse and then on foot to see dawn from the lip of the crater and watch as the first piercing rays of the sun reveal the crater walls, the other peaks, and the mist over the 'Sand Sea', all in changing colours as the sun climbs. Mount Semeru (sometimes called Mahameru) at 3,676 metres (12,060 feet) is Java's highest and perhaps steepest peak and is far from an easy climb, and indeed not to be advised since it is still quite active. Below about 3,000 metres (9,800 feet), however, the belt of lower montane forest is very fine. Around the main crater itself, as elsewhere in East Java above 1,400 metres (4,600 feet), the most typical tree is the fine-leaved Cemara which is a pioneer on fresh volcanic ash. Where fires are frequent, however, it cannot survive and is gradually replaced by grasslands. The famous Sand Sea of porous sand and ash around Bromo looks more like a desolate steppe or moonscape and is nearly devoid of vegetation except, in more sheltered areas, some hardy binding grasses. The park has little in the way of mammals because of past hunting pressure, but patient birdwatching has revealed an interesting and diverse community, including birds of prey which are generally very rare on Java.

View across the larger Tengger and smaller Bromo craters to the smoking cone of Mount Semeru, at 3,676 metres (12,060 feet), the highest mountain in Java.

ABOVE Sunrise seen from the lip of the Bromo crater looking across the 'Sea of Sand' to the walls of the Tengger crater.

BELOW Early morning mist rising from the 'Sea of Sand'; *Casuarina junghuhniana* trees clothe the slopes.

RIGHT Five volcanoes in a single view: the crater of Mount Tengger, the flat-topped Mount Batok, the squat, mist-filled Mount Bromo, the larger Mount Kursi, and the active Mount Semeru behind.

ABOVE View into the steaming Bromo crater, a reminder that this seemingly benign volcano may just be resting.

LEFT Looking up to the outer wall of Bromo crater showing the hardy grasses, sedges and shrubs growing at the edge of the 'Sea of Sand'. Few plants can tolerate this hostile environment.

Baluran National Park

Baluran (250 square kilometres or nearly 100 square miles) at the north-east tip of Java was one of Indonesia's first national parks because of its accessibility, ease of viewing large wildlife, the striking landscape and its dry climate. The main vegetation of interest is wooded savannah, including a tree endemic to the park, *Erythrina euodophylla*. Baluran has relatively large populations of wildlife such as Peafowl, Green Junglefowl, Leopard, Banteng and Timor Deer. Visitor facilities, such as look-out towers, are well developed, but the park is experiencing some major problems. For example, half the park is barely used by the large species because annual fires and domestic livestock have reduced the vegetation to little more than poor agricultural land. Perhaps even more serious, a species of tree, *Acacia nilotica*, was introduced to the edge of the park about twenty years ago as a 'green fence', but it has spread out of control and is on the verge of taking over the grassland near the viewing towers that was an important feeding ground for the large herbivores. In addition, the exotic weed *Lantana camara* is also spreading widely and, because it is toxic to herbivores, it is reducing the area of grazing available. Lastly, the large numbers of feral Water Buffalo may be competing with the indigenous wildlife and are a serious potential source of disease that could spread to the Banteng.

View across the park to the extinct volcano of Mount Baluran (1,556 metres or 5,105 feet).

ABOVE There are several hundred feral Water Buffalo (*Bos bubalis*) within the park. They are a tourist attraction but may be competing with the Banteng and other indigenous wildlife, and are a potential source of disease.

ABOVE RIGHT Fine bull Banteng (*Bos javanicus*) with a young cow.

RIGHT A Silvered Leaf Monkey (*Trachypithecus auratus*) leaping between trees. This is the most abundant primate in the park; a red form is also found.

BELOW The splendid Green Peafowl (*Pavo muticus*) is quite common in the park, and can be seen scraping in the ground for seeds, termites, grasshoppers and small reptiles. Its morning trumpeting calls echo through the woodlands, particularly during the breeding season from August to October, when the females lay three or four brownish eggs in a grassy hollow.

BELOW RIGHT The Pangolin (*Manis javanica*) is one of Indonesia's stranger mammals, with scales instead of fur. It feeds on termites by breaking open their nests with its strong claws and then licking up the termites and their larvae with its long sticky tongue.

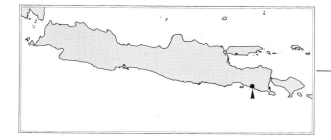

Meru Betiri National Park

Meru Betiri National Park covers 500 square kilometres (190 square miles) and is virtually the last area of lowland rain forest remaining in Java. It was formed as a reserve in 1972 to safeguard the last population of the Javan subspecies of the Tiger, but this proved to be too little action too late, and the Tiger is probably now extinct. Even so, the area is a fine expanse of forest and coast which deserves to be appreciated by as many people as the park itself can stand. The area is hilly and the highest point, Mount Betiri, is 1,223 metres (4,012 feet) above sea level. The steepness of the hills makes most of the area inaccessible and completely unsuitable for sustainable agriculture (though this is often not enough to deter a lone farmer with only a short-term perspective). Much of the south of the park comprises rocky cliffs up to 300 metres (nearly 1,000 feet) high, but there are also some sandy beaches up to 3 kilometres (2 miles) long, on some of which females of four species of turtles (predominantly Greens) pull themselves up to lay eggs. The nests are protected by park staff until the eggs have hatched. There are about 75 to 100 Banteng and about 40 Leopard within the park, and Rusa Deer introduced in 1978 as prey for the Leopard appear to have become established. The park is surrounded by rubber and coffee plantations and, although the human population density is low compared with elsewhere in Java, there are still about 100,000 people living within easy walking distance, including some in illegal settlements, and the resultant loss of forest products, such as bamboo, rattan and firewood, is extremely damaging.

Forested slopes within the park reach down to the clear blue sea on the south coast.

ABOVE Forested hills – note the deciduous species which have shed their leaves.

LEFT Scaly fruits of the *Daemonorops melanochaetes* rattan palm. Despite prohibitions to the contrary, people come and cut rattan for sale to the furniture industry in distant towns such as Semarang on the north coast.

BELOW Buds and a decaying flower of *Rafflesia patma*. Collectors seek out these flowers because of their uses in traditional medicines, selling them to factories near Surakarta (Solo).

TOP Adult female Green Turtle (*Chelonia mydas*) covering a nest hole in which her eggs are laid at Sukamade Beach. These eggs will be protected under a management scheme and the young given a slightly greater chance of surviving to adulthood.

ABOVE Young Sand Crab (*Oxypode* sp.) the adults of which normally occupy small burrows near the high-water mark. These crabs feed mainly on organic material but sometimes also predate on small crustaceans such as sand hoppers and other crabs.

ABOVE RIGHT The loose folds of skin along the belly of the Flying Lizard (*Draco volans*) have rib extensions within them and can be spread when the animal leaps off a tree. This enables it to glide to the next tree, using its tail as a rudder.

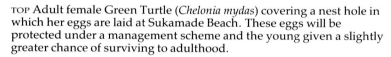

RIGHT The Green Jungle Fowl (*Gallus varius*) is restricted to Java and the islands of Flores and Sumba. It favours open ground, being frequently seen in teak plantations. This bird also visits the outskirts of villages, but does not appear to interbreed with domestic chickens, which were derived initially from the Red Junglefowl (*Gallus gallus*).

Bali

Most visitors to Bali do not see the wild side of this relatively small island. The popular tourist areas are concentrated on the beaches and the fertile plain in the south. The forest areas are mainly in the mountains in the northern half of the island and in the west. Good walks into the cool and atmospheric mountain forests can be enjoyed, starting from the Botanic Gardens at Candikuning near Lake Bratan (home to Bali's endemic *Rasbora* fish), from Lake Tamblingan, or from Pura Luhur, and the adventurous can strike north from the western coast road along dirt roads which generally peter out close to the forest edge. It is possible to walk across this narrow western arm to the dry north coast, though it may be difficult to find a guide. At the western tip of the island is Bali Barat National Park with a complex of habitats encompassing wet forests in the south and dry forests to the north. The park's most famous inhabitant is the Bali Starling, which can be seen on trips arranged by the park staff. Also popular is the coral-fringed island of Menjangan; indeed, many visitors come to Bali simply to experience diving around the reefs. To the east of Bali there are dramatic lava fields around the volcanoes of Batur and Agung, and around the bare cone of the latter botanically interesting Cemara forests can be enjoyed, starting from the 'Mother Temple' of Besakih.

ABOVE Southern Bali and its off-lying islands have some dramatic cliffs below which dolphins, turtles and schools of large fish can be seen. The narrow cliff ledges provide nesting sites for the elegant White-tailed Tropicbird (*Phaeton lepturus*).

OPPOSITE PAGE Rivers have cut steep gorges through the volcanic rock of central Bali, such as here near Gunung Kawi.

RIGHT Indonesia is both the world's largest producer and importer of cloves, which are used in *kretek* cigarettes, and plantations such as this can be very profitable.

The Bat Cave temple in eastern Bali is inhabited by vast numbers of pollen-eating and nectar-drinking Cave Fruit Bats (*Eonycteris spelaea*) which, being protected there, allow visitors to approach them.

Long-tailed Macaques (*Macaca fascicularis*) are a common sight along certain roads and at a number of temples, where they pester visitors for food.

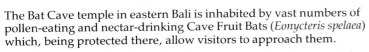

The village of Petulu in central Bali won a national environmental award for encouraging a population of some 7,000 Cattle Egrets (*Bubulcus ibis*) to roost safely along its main street.

Iridescent damselfly (*Vestalis luctuosa*) by a river at 1,000 metres (3,300 feet) in central Bali.

The dragonfly *Crocothemis servilia* can quite often be seen near the forest edge and around the beautiful Balinese rice terraces.

Scaevola taccada is a small herb that grows on unstable limestone cliffs in southern Bali.

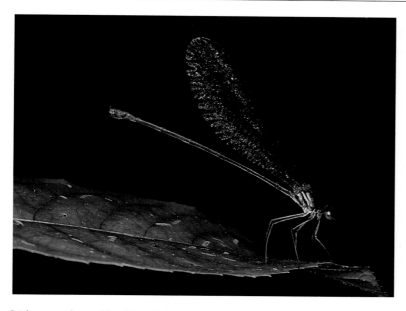

Long-winged fruit of the tall forest tree *Dipterocarpus hasseltii*, which dominates the 10-hectare (25-acre) holy forest in the Sangeh Nature Reserve.

The forest fig, *Ficus aurantiaca*.

Fronds of the fern *Bolbitis heteroclita*.

Focus on Sulawesi

The shape of Sulawesi has been described in many ways: like a wind-blown orchid, a drunken spider, a scarecrow in a hurricane, and a child's letter 'K'. It is characterized by four long peninsulas which give it the longest coastline relative to land area of any of the islands in Indonesia. Nowhere is further than 90 kilometres (55 miles) from the coast and most places are within just 50 kilometres (30 miles). The coasts are wrapped either in fertile mudflats backed by mangrove forests in relatively peaceful areas, or beautiful beaches of white sand in the south, and in black, volcanic sand in the north where the sea action is naturally more rough or where man has removed the protective mangrove fringe. Coral reefs abound, with major areas found just west of the main town of Ujung Pandang, around the increasingly famous Bunaken Island near Menado, around the Togian Islands at the mouth of Tomini Bay, in the much more remote islands of the Tukangbesi group in the south-east, and in the world's third largest atoll of Taka Bone Rate just north of Flores.

Inland, Sulawesi is largely mountainous with 20 per cent of the land above 1,000 metres (3,280 feet). The highest peaks are in the central hub of the island, the tallest being Rantemario at 3,455 metres (11,335 feet) which is reasonably accessible from the main road to the popular tourist destination of Toraja at the top of the south-west arm.

The south of the south-west peninsula has a number of extinct volcanoes – the largest being Lompobatang behind Ujung Pandang – which have spread fertile debris over the land, resulting in a very productive agricultural area similar to parts of Java. Up in the north on the mainland and on the smaller islands are eleven volcanoes, some of them singularly wakeful and at least one of which is generally spewing out ash. Occasionally, as in 1983 on Una-Una Island south of the northern arm, very dramatic eruptions occur – although through an extensive prediction network, people are usually evacuated in time. As recently as 1966, however, 7,300 people were killed when Awu volcano on Siau Island erupted.

Sulawesi and its offshore islands cover an area of 186,145 square kilometres (71,870 square miles) and its longest axis spans 1,805 kilometres (1,122 miles) from the Satengar Islands in the extreme south-west, just north of Sumbawa, and Miangas Island just south of the Philippine island of Mindanao; this is further apart than the northern and southern tips of Sumatra. There are four provinces: North Sulawesi (19,023 square kilometres or 7,345 square miles) with its active volcanoes; mountainous Central Sulawesi (69,726 square kilometres or 26,921 square miles); South Sulawesi (72,781 square kilometres or 28,101 square miles) with extensive ricefields, lakes, Torajaland, and rapid industrial growth in the south; and South-east Sulawesi (27,686 square kilometres or 10,690 square miles) little explored and seldom visited by travellers.

Sulawesi has some 12.7 million people, a total which is increasing by about 2 per cent annually. More than half of the population live in the fertile plains and valleys of South Sulawesi, and another million or so live in the capital of North Sulawesi, Manado. The dominant group are the Buginese, well known for travelling around the archipelago and beyond to establish coastal settlements and trading posts. The number and make-up of the different Sulawesi ethnic groups is extremely complex. There are many languages and dialects (Central Sulawesi alone has fifty ways of saying 'no'), many of them very distinct.

As with many of the islands in eastern Indonesia, the climate of Sulawesi shows distinct geographical variation. In the central massif and at the northern tip the climate is wet all year round, whereas around the towns of Palu, Gorontalo, the southern tip of the south-west arm and Ujung Pandang the climate is markedly seasonal. This is most acute around Palu where the lowest rainfall in Indonesia is recorded.

About 60 per cent of Sulawesi is still under some form of forest cover, most of it in the hills and mountains. Wet field rice accounts for only 5 per cent of the land area, and this is concentrated in southern South Sulawesi, which is one of the few provinces outside Java and Bali to produce more rice than it consumes. Over one-fifth of the land is covered in bush, scrub and shifting cultivation, a form of agriculture which is environmentally appropriate only where the human population density is low.

Industrial activities are concentrated around Ujung Pandang where there are cement works, major coastal and inland fisheries, and many light and heavy industries. Near Lake Matano there is a major nickel mine which has transformed the small, quiet town of Soroako into a major centre of activity in less than twenty years.

The wildest areas in Sulawesi are in the central massif, but even in the narrow northern arm some locations can be reached only by three days of walking and hauling boats over rapids. The largest conservation area is Dumoga-Bone National Park near Gorontalo, followed by Lore Lindu National Park south of Palu. The Tangkoko Reserve east of Manado is small but it has exciting animals and plants and is extremely accessible. Also near Manado is the marine park around Bunaken Island which provides some of the best and most rewarding coral viewing in the archipelago.

Landscape across Torajaland in mountainous South Sulawesi. The style of the houses at the end of the rainbow is typical of this area.

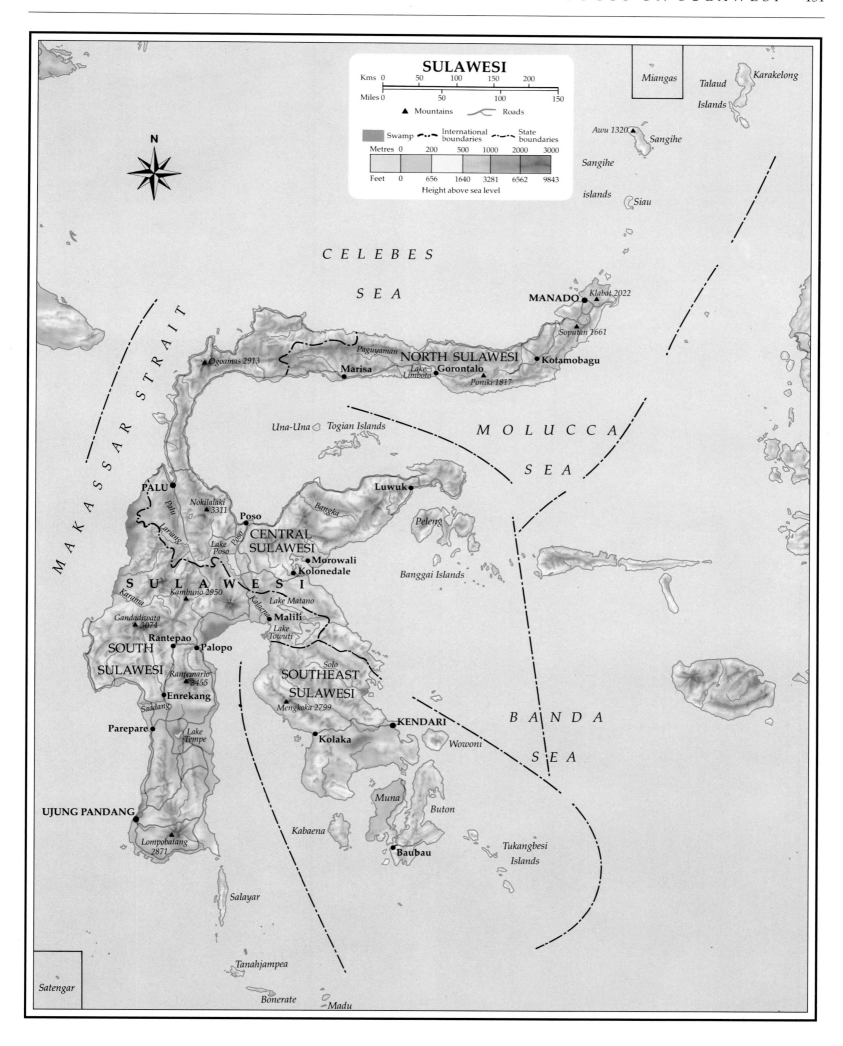

SULAWESI

Kms 0   50   100   150   200

Miles 0   50   100   150

▲ Mountains   ⌇ Roads

▨ Swamp   ·—·—· International boundaries   ·—·· State boundaries

Metres 0   200   500   1000   2000   3000

Feet 0   656   1640   3281   6562   9843

Height above sea level

N

Miangas

Talaud *Karakelong*

Islands

Awu 1320 ▲ *Sangihe*

Sangihe

islands ◌*Siau*

C E L E B E S

S E A

MANADO ● *Klabat 2022* ▲

▲ *Soputan 1661*

Paguyaman **NORTH SULAWESI** ● **Kotamobagu**

▲ *Ogoamas 2913* *Lake* **Gorontalo**

Marisa *Limboto* ▲ *Poniki 1817*

M O L U C C A

Una-Una *Togian Islands* S E A

M A K A S S A R   S T R A I T

PALU ●

Palu *Nokilalaki* *Bangka* **Luwuk** ●

▲ *3311*

Lariang **Poso** ● *Peleng*

Lake **CENTRAL**

Poso **SULAWESI** *Banggai Islands*

Karama **S U L A W E S I** ● **Morowali**

▲ *Kambuno 2950* ● **Kolonedale**

Gandadiwata *Lake Matano*

▲ *3074* ● **Malili**

Rantepao ● *Lake*

SOUTH ● **Palopo** *Towuti*

SULAWESI *Rantemario* *Solo*

▲ *3455* **SOUTHEAST**

● **Enrekang** **SULAWESI**

Saadang *Mengkoka 2799*

Parepare ● *Lake* **KENDARI** ●

Tempe **Kolaka** ● B A N D A

Wowoni

S E A

UJUNG PANDANG ●

Lompobatang *Muna*

2871 *Buton*

Kabaena

Salayar ● **Baubau** *Tukangbesi*

Islands

Tanahjampea

Satengar *Bonerate* *Madu*

Tangkoko Reserve

Tangkoko is a small reserve at the northern tip of the Minahasa Peninsula in North Sulawesi, which covers only 87 square kilometres (34 square miles). It is quite easily reached from the North Sulawesi capital of Manado, just 40 kilometres (25 miles) away, via the port of Bitung, and it encompasses three volcanoes, Tangkoko, Batuangus and the double-peaked Dua Saudara, and ranges from a marine and coastal zone up to the top of Dua Saudara at 1,351 metres (4,432 feet). The last major eruption was in 1839 and the soil comprises small pieces of clinker ash. The lowland rain forest which cloaks most of the reserve is thus quite young and is highly productive, supporting many frugivores and other animals. Indeed, the density of the Black Macaque is higher than in any other area studied. Other endemic species include the Anoa, the Sulawesi Pig, Tarsier, Sulawesi Civet, both species of cuscus, and a variety of rats. The Babirusa appears to be absent. Among the birds the large Sulawesi Hornbill is extremely common and the Maleo has several communal nesting sites. Thus, despite its small size, the reserve is very important for the conservation of many of Sulawesi's endemic species. There is continual pressure on the boundaries of the reserve and occasional quasi-accidental fires make inroads into the forests. The isolation of this forest area may result in the eventual local extinction of the Anoa and Maleo, but proper management will help to delay this, particularly for the Maleo.

Approaching Tangkoko Reserve with a coconut plantation in the foreground.

ABOVE Fig tree (*Ficus* sp.) with ripening fruits growing directly out of the trunk. It is thought that this is an adaptation favouring dispersal by fruit bats which can reach the fruit without getting caught in the branches.

ABOVE A 'strangling' fig (*Ficus* sp.) with huge roots growing down from the parent plant in the canopy.

BELOW The Round-leaved Fan Palm (*Livistona rotundifolia*) is characteristic of many Sulawesi lowland forests, where it may grow in abundance. Its leaves can reach 130 centimetres (50 inches) in diameter and the stems have very sharp spines.

BELOW These plants eventually kill their hosts but not by strangling them, as was thought. Instead the fig shades out its host tree which eventually dies, leaving a hollow core used by bats, snakes and other animals.

LEFT Black Macaques (*Macaca nigra*) live at a higher density here than any other primate in a natural habitat in Asia: about 300 per square kilometre. This is possible because of the very high density of suitable fruit trees.

ABOVE The flowers of the understorey shrub *Clerodendrum* sp. When the flowers first open, as shown here, the anthers project, providing a landing place for insects. Later these roll up and the style takes their place.

LEFT Red-knobbed Hornbills (*Rhyticeros cassidix*) are the larger of the two hornbill species endemic to Sulawesi. These important seed dispersers live at very high densities in the reserve.

ABOVE The Dwarf Cuscus (*Strigocuscus celebensis*) is nocturnal and secretive in its habits, so it is only rarely seen. It eats fruit and perhaps some animal prey when it can be caught.

ABOVE The Flying Lizard (*Draco volans*) can be seen in forest, rural and even urban environments. A patient observer will be rewarded by watching it glide from tree to tree, scuttling upwards as soon as it lands. BELOW Eggs are laid in leaf mould and hatch after a few weeks.

RIGHT Spectral Tarsiers (*Tarsier spectrum*) are the smallest primates in Indonesia. They live in small family groups but hunt separately at night for large insects such as grasshoppers, stick insects and beetles which they grab with their long-fingered hands, and crunch in a satisfying manner.

Bunaken Island Reserve

Just offshore from Manado, capital of North Sulawesi, is a group of islands. The most striking is the beautiful cone-shaped extinct volcano of Manado Tua. To the east is a low-lying elongate island called Bunaken which has become world famous in the last five to ten years because of the excellent quality of its coral, particularly the sheer-sided fringing reefs. Credit for its popularity and the accompanying environmental concern must go to the Nusantara Diving Club, just south of Manado, which in 1985 (one year before the reserve was established) received one of the Kalpataru Environment Day awards presented by the President at the Jakarta Palace. This club brought the beauty and economic role of the reefs to the attention of the public and the local authorities. It even convinced the port authorities not to allow large ships to pass the straits between Manado and Bunaken, where most of the best reefs are situated, but rather to encourage the use of the port at Bitung on the other side of the peninsula. The sandy beaches of Bunaken are used by turtles for breeding, although it is not yet known how important they are. There are plans to develop a major tourist centre based on the coral in Manado Bay, and it must be hoped that too much or uncontrolled attention does not destroy the very resource that attracts visitors.

The Bunaken Island Reserve comprises four islands, the most striking of which is the ancient volcano of Manadotua. Bunaken Island itself is on the right.

RIGHT A typical reef to the south of Bunaken Island with myriads of colourful, small fishes.

RIGHT Shrimpfish or Razorfish (*Aeoliscus strigatus*) live in shallow waters around coral reefs. They eat plankton and are generally found in close schools, swimming head down and moving slowly across the reef. If alarmed, they will swim away in a more normal horizontal position. These fishes sometimes camouflage themselves among the spines of a sea urchin.

BELOW RIGHT Scorpionfish (*Pterois antennata*) are generally solitary and swim slowly over the reef, preying on small fishes, crabs and prawns. Their large fins have sharp spines with poison glands. Because of their camouflage and sluggish movements, they are a threat to waders and swimmers and a wound caused by a spine is extremely painful and sometimes dangerous.

BELOW CENTRE A large crinoid or featherstar with a White-spotted Butterfly Fish (*Chaetodon kleini*) below. This fish is common on patch reefs interspersed with sand, usually being found in small groups. It is said to feed on algae and small invertebrates picked up from the coral or sand.

BELOW Nudibranch sea slug *Chromodoris*.

ABOVE The Clown Anemone Fish (*Amphiprion ocellaris*) is the most popular anemone fish among aquarium enthusiasts, but only under very specific conditions does it thrive or breed. In the wild it is found in calm, shallow water where an adult pair and two or three young occupy an anemone. A mucus layer on the fish's skin protects it from the anemone's stings.

BELOW A poisonous Scorpionfish (*Scorpaenopsis* sp.) resting across the top of a barrel sponge.

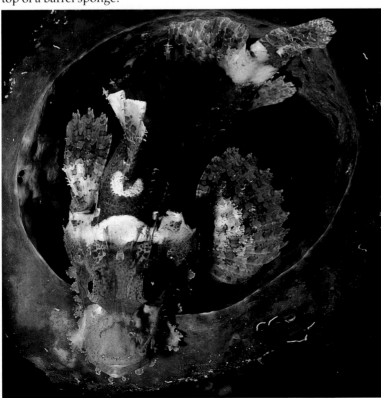

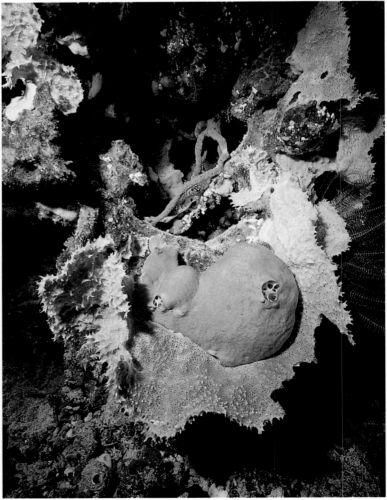

ABOVE A *Gorgonia* sea fan with a large crinoid or featherstar perched on one edge, its arms extending into the main sea current in order to maximize the number of the small animals it catches.

ABOVE RIGHT *Gorgonia* sea fan, a relative of the soft corals.

RIGHT Starfish (possibly *Protoreaster nodosus*) found in the seagrass meadows adjoining reefs.

RIGHT Stony coral *Montipora*.

OPPOSITE PAGE A colourful community of at least five types of sponge (pink, orange, lilac, grey and yellow), small blue sea squirts, and a featherstar with a small brittlestar above it.

Dumoga-Bone National Park

Dumoga-Bone National Park is about 3,000 square kilometres (1,160 square miles) in the middle of the long straggling arm of North Sulawesi. It is the watershed to two major rivers, the Dumoga in the east flowing towards Kotamobagu, and the Bone in the west flowing towards Gorontalo. Access is either from Gorontalo or Manado, both of which are served by airports. The lowest areas are only 200 metres (660 feet) above sea level while the highest reach nearly 2,000 metres (6,600 feet) at the tallest peak, Mount Ganbuta. Most of the park comprises complex, long, craggy ridges, many of them of ancient volcanic origin, although some hot springs and dramatic precipitated limestone 'frozen waterfalls' can still be found. More than half of the park is covered in rich lowland rain forest with abundant fan palms and, along the rivers, the native *Eucalyptus* gum tree. Almost all Sulawesi's endemic mammals and birds are present, including good populations of the Maleo for which at least ten communal nesting sites are known, mostly associated with volcanic vents.

This park has been the focus of research by individual scientists, as well as by a major British entomological expedition, Project Wallace, in 1985. Its management has been supported by World Bank funds because it was recognized that the success of the irrigation scheme in the Dumoga Valley, also funded by the Bank, depended on a protected watershed. Even so, the park faces considerable threats from illegal forest clearance, hunting (particularly of the Babirusa which is already uncommon in the eastern half), road building, and from the possibility of copper mining in the western half.

Just after dawn the mist rises off the forest at the edge of the park.

ABOVE The Sulawesi Flying Fox (*Acerodon celebensis*) is one of six fruit-eating bats endemic to Sulawesi.

LEFT Maleo birds (*Macrocephalon maleo*) lay their eggs in holes which they dig near volcanic vents or in beach sand. The adults do not sit on the eggs but rely on the heat of the sun-baked sand or volcanic zone to incubate them.
BELOW The Dumoga Valley used to be covered by freshwater swamp forest, but this has now been converted to fertile ricefields. Large populations of Cattle Egrets (*Bulbulcus ibis*) and similar birds can be seen, but they are threatened by the increasing use of pesticides.

BELOW The Timor Deer (*Cervus timorensis*) was introduced to Sulawesi centuries ago but is now found throughout the island.
BOTTOM Babirusa (*Babyrousa babyrussa*) are uncommon or absent in the east of the park, probably as a result of hunting, but are still found in the west.

Lore Lindu National Park

Lore Lindu lies south-east of Palu, the capital of Central Sulawesi, and covers an area of about 2,300 square kilometres (890 square miles). It ranges from valley bottoms at about 200 metres (660 feet) to the peak of Mount Rorakatimbu at 2,610 metres (8,563 feet), and most of the area is over 1,000 metres (above 3,000 feet). The land is generally steep and dissected, particularly in the north, where it is higher. Lake Lindu lies in the west and is most interesting, partly because it is the only site in Indonesia where the parasite responsible for the debilitating disease schistosomiasis occurs, and partly because at least one endemic fish and several endemic molluscs occur in its waters. About 90 per cent of the park is covered in montane forest with various oaks, chestnuts, laurels and conifers predominating. Most of Sulawesi's mammals and birds can be found within the borders of this exceptional park even though the extent of lowland forest is minimal, but it is possible that certain species have a very limited distribution because of this. Within or very close to the park, in the Besoa, Napu and Bada Valleys, are some of the most impressive megalithic stones to be found in Indonesia, in the form of figures or sarcophagi. Efforts are made to maintain the integrity of the forests within and around Lore Lindu in the face of illegal logging for ebony and of environmentally inappropriate forms of shifting agriculture, and increasing attention is being paid to the park by tour agencies who exploit the local mode of transport, riding on horseback, as an attraction. Meanwhile, proposals have been made to make Lake Lindu the site of a major hydroelectric project, the environmental impacts of which would possibly be quite severe.

RIGHT The only gum tree found in rain forest is *Eucalyptus deglupta*, which occurs in Sulawesi near rivers. Its peeling trunk and imposing size make it unmistakable.

OPPOSITE PAGE The majestic *Pigafetta filaris* palm of Sulawesi and the Moluccas is unique among South-east Asian palms, being adapted to growth in disturbed habitats: it grows fast, tolerates bright sunlight and produces large numbers of fruit.

BELOW View towards the mountains of the park showing the contrast between the grassy plains and the forested slopes.

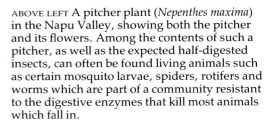

ABOVE LEFT A pitcher plant (*Nepenthes maxima*) in the Napu Valley, showing both the pitcher and its flowers. Among the contents of such a pitcher, as well as the expected half-digested insects, can often be found living animals such as certain mosquito larvae, spiders, rotifers and worms which are part of a community resistant to the digestive enzymes that kill most animals which fall in.

ABOVE CENTRE This rhododendron (*Rhododendron javanicum* var. *schadenbergii*) occurs in Sulawesi and the Philippines where it is found in rocky areas and near volcanic craters, as well as on trees in primary and secondary forests. Its flowers, which can be found year round, are said to be edible.

ABOVE *Phaius tankervillae* is a widespread, terrestrial orchid with metre- (3-foot) long leaves and 12-centimetre (5-inch) wide flowers. The genus used to be popular as conservatory plants in Europe, and this species was named from a specimen cultivated in England in the mid-eighteenth century.

LEFT ABOVE Moss and liverwort on the moist trunk of a lowland forest tree.

LEFT A pair of ancient megaliths near the village of Besoa in the Napu Valley whose meaning and creators are unknown. Although megaliths from similar 'Bronze Age' cultures are known from elsewhere in Indonesia, there was no single megalithic culture, and there is no proven link with similar statues elsewhere in Asia.

ABOVE A pair (female below) of what appears to be a new species of the forest grasshopper *Chitaura*.

ABOVE RIGHT Blume's Swallowtail (*Papilio blumei*), a species endemic to Sulawesi.

RIGHT These butterflies are probably Irene's Grass Yellow (*Eurema irena*), a species endemic to Sulawesi.

BELOW RIGHT The endemic forest rat *Maxomys hellwandii*.

BELOW CENTRE Sulawesi's endemic toad, *Bufo celebensis*, in full song.

BELOW Small black sweat bees and honey bees on sand sucking up moisture, perhaps where an animal has urinated, in order to collect salts.

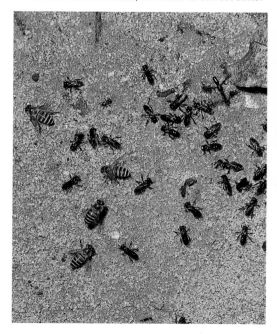

Focus on the Moluccas

The Moluccas is a province of about a thousand scattered islands, spanning 1,300 kilometres (800 miles). They range in size from Seram and Halmahera, which are about 20,000 square kilometres (7,720 square miles) in area, to many which are so small they are not shown on most maps. The average annual rainfall of the Moluccas is 2,370 millimetres (90 inches), but the distribution of the rainfall varies throughout the province from year round in Seram and northern Halmahera to markedly seasonal in southern Halmahera, Obi, north-east Buru and the smaller islands to the south.

There are two lines of volcanic islands, the Wetar-Damar-Banda arc in the south, and the Ternate-Tidore-Makian arc to the north. The other northern islands are largely mountains of granite overlain by sedimentary rocks. In the south the non-volcanic islands are mainly made of sedimentary limestones, and the Aru and Kai Islands are up-raised coral reefs. With the exception of the Aru Islands which sit on the Sahul Plate dominated by New Guinea, all the islands in the Moluccas are oceanic, that is, they have never been connected with larger land masses. Islands which have split from larger land masses will have a flora and fauna which reflects this, but all the plants and animals on oceanic islands have arrived over the water. Hence the more mobile animals and the plants with readily dispersed seeds will be better represented. In the Moluccas this has resulted in a large and varied bird fauna (though no single island has very many species) but a relatively poor flora and mammal fauna, and fish fauna in the rivers comprising species with tolerance to salt water.

The Moluccas is a single province of 74,505 square kilometres (28,766 square miles) of land set in a huge area of sea of nearly 1,000,000 square kilometres (386,000 square miles). The capital is Ambon, a bustling town on the rather small island of the same name just south of Seram. The only other town of any size is Ternate, also on an island with the same name. The province has a population of about 1.8 million, comprising a rich mixture of Malays from the west and Melanesians from the east. In earlier times almost all of them were dependent on the sea and its resources, although this is now changing as other means of earning a livelihood become available. A few tribal groups remain, such as the Naulu in central Seram and the Togutil in eastern Halmahera.

The Moluccan islands are the fabled 'spice islands' of the East, over which many battles were fought and lives lost. During the Middle Ages the world's whole supply of nutmeg and mace (*Myristica fragrans*), cloves (*Eugenia aromatica*), candlenuts (*Aleurites moluccana*) and kenari (*Canarium commune*) came from this province, and this at a time when they were literally worth their weight in gold. Today Indonesia is still the world's major producer of cloves, but the contribution of the Moluccas is no longer as significant. The main staple is rice but in some areas, such as Seram, sago is of major local importance. The timber industry is important on the larger islands to the north, and even on the smaller islands around them, which have not been thoroughly surveyed biologically, the remaining untouched forests have been parcelled out as logging concessions. Another main industry is related to fish and shrimps which are frozen in various states and exported to Japan and other countries. In addition, pearls are an important product from around the Aru Islands. Traditional fishing methods and vessels have to a certain extent been replaced by large industrial ships, many of them from outside Indonesia, but the local fleets are still significant in the domestic markets. The Moluccan islands have little mineral wealth, but gold has been discovered on Wetar, which may bring major changes to the island.

The name Moluccas speaks of small islands with deserted, white-sand beaches sloping down to calm azure seas ebbing and

View from the lotus-fringed Lake Laguna in the south of Ternate towards the active Gamalama Volcano (1,721 metres or 5,646 feet) in the centre of the island. This volcano has erupted seventy times in the last 500 years, one reason why almost all the villages are in the coastal fringe.

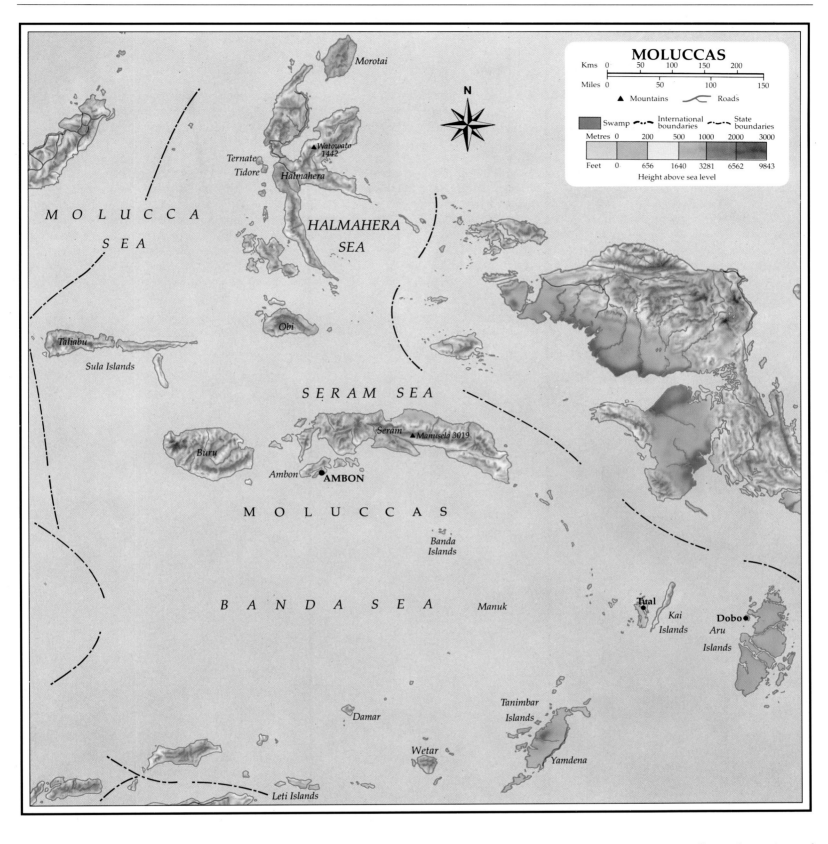

flowing over colourful and productive coral reefs. This is one type of wildness, best exemplified perhaps by the Banda and Kai island groups. The wild forest areas of the Moluccas are the remote and little-known forests of Halmahera, Sula, Seram and Buru. Of all these islands Halmahera has by far the greatest number of species, the highest level of endemism, as well as the widest range of climates and landforms, and is thus the highest priority for conservation activities in the province. Although this has been realized for at least a decade, virtually no conservation moves have yet been made on the island. The major gazetted conservation area in the province is the Manusela National Park in central Seram, which has an exceptionally wide variety of animal and plant life.

These widely scattered islands are difficult to travel among and so the province has attracted less wildlife research than other parts of the country, and there is still much to learn. For example, in 1990 a small team of ornithologists visited Wetar, the first such expedition since 1911, and they found many previously unrecorded species. Manusela National Park stands as an exception to this rule because it has received a great deal of attention since it became one of the research sites of Operation Raleigh, the British-led international expedition.

Banda Islands

The nine, small, volcanic Banda islands dominated by the perfect cone of Mount Api volcano (676 metres or 2,218 feet) found fame as the only source of nutmeg and mace until the beginning of the nineteenth century. During this time their location was a closely guarded secret known only to a few Arab traders before the Portuguese found the islands in 1511. (Both spices come from the same fruit – the nutmeg being the hard seed and the mace being a red, lace-like covering between the seed and the thick skin.) At that time, the islands were an important entrepôt, trading also in cloves from Ternate and Tidore, bird of paradise feathers from the east, and slaves. Later the Dutch and British came to build forts and many lives were lost in the struggle for monopoly and large profits. Banda was also used by the Dutch to hold exiled nationalists, including

Mohammad Hatta, the first Vice-President of independent Indonesia. Most visitors coming to these islands now do so as much for the exceptional crystal-clear underwater scenery as for the beautiful islands above the sea. The Banda Islands Marine Reserve covers effectively the entire archipelago and both scuba diving and snorkelling are very rewarding. In many areas the situation has not changed since the mid-nineteenth century when Alfred Russel Wallace was able to write of these waters 'even the minutest objects are plainly seen on the volcanic sand at a depth of six or seven fathoms'. Suanggi (Ghost) Island can be visited by boat and is interesting for its nesting seabirds, but rats and human disturbance are making their respective impacts. Sharks are said to patrol around its craggy coast.

ABOVE View of Mount Api volcano, 676 metres (2,218 feet), and the small Pisang Island showing the lava stream from the 1988 eruption.

OPPOSITE PAGE The smouldering, lifeless summit of Mount Api. During its last eruption, house-sized burning rocks were thrown into the air and a swathe of lava flowed down to the west, smothering everything in its path and charring living plants on either side.

LEFT Trees near the path of the lava were killed by the intense heat, but some have sprouted from their bases, and trees on hillocks have begun to regrow. Seeds will be dispersed from these and so vegetation will develop once again.

BELOW Nutmeg and mace are the seed and red seed-skin respectively of the *Myristica fragrans* tree. The nine, small volcanic Banda Islands were the only source of these spices until the beginning of the nineteenth century, and their location was a closely guarded secret for hundreds of years.

ABOVE The Nicobar Pigeon (*Caloenas nicobarica*) is immediately recognizable by its white tail and long neck feathers. It generally roosts and nests on small islands, flying to the mainland or neighbouring larger islands to feed. It is found from the Andamans and Nicobars north of Sumatra to the Solomons in Melanesia.

ABOVE The elegant Imperial Pigeon (*Ducula concinna*) lives only on small islands from Buru to New Guinea. It feeds on ripe nutmegs, which are adapted to dispersal by pigeons by hanging from trees part open, exposing the bright red aril around the seed.

OPPOSITE PAGE The crater of Mount Api.

Kai Islands

The Kai Islands are about as close as one will find to the perfect tropical islands. Nestled beneath the 'neck' of the Bird's Head Peninsula of Irian Jaya, they lie in the placid multi-hued blue waters of some of the furthest reaches of the archipelago. The two main islands are very different; the smaller, Kai Kecil, has a tortuously curved coastline and is largely flat, heavily populated and converted to different forms of agriculture. The larger, Kai Besar, is forested and hilly, with a sparse human population concentrated around the edge. Its forests are home to no fewer than five species of birds, one species of bat, two species of lizards and one species of snake not known from anywhere else in the world. Hardwood trees from the forests were used to make magnificent boats, the fine construction of which caused

Alfred Russel Wallace to marvel when he passed through Kai in 1857. These traditions are still practised on Kai Besar. Although the islands are remote from the rest of Indonesia, the same pressures experienced elsewhere – from the cutting of trees and clearance of land for small farms – threaten the conservation of the area. No reserve has yet been set up here, and it is hoped that it will not be too long before one is and its regulations are enforced.

A remote, coconut-fringed beach on Kai Kecil Island.

ABOVE Coconuts are an important crop throughout Indonesia and the tree has been well named 'the tree of life'. Although widespread in the tropics, it probably originated in eastern Indonesia or Melanesia. Various parts of the coconut are used for food, drink, oil, medicine, fibre, timber, thatch, mats, fuel and utensils.

ABOVE Natural beach forest on Kai Kecil Island.

BELOW The Spotted Cuscus (*Spilocuscus maculatus*) is found in Kai as well as in New Guinea and north Australia.

BELOW Screwpines (*Pandanus* sp.) are commonly found among beach vegetation.

Focus on the Lesser Sundas

The Lesser Sundas is the term given to the relatively small islands in the south of the Indonesian archipelago, separate from the 'Greater Sundas' of Java, Sumatra and Borneo. From a geographical point of view, Bali is a member of this group but is separated on biogeographical grounds because it is much more closely related to Java. The Lesser Sundas, or Nusa Tenggara as they are known within Indonesia, are spread across 900 kilometres (560 miles) from Lombok in the west, through Sumbawa, Flores and Sumba, to Timor in the east. The inner arc of islands, from Lombok to Alor are of volcanic origin, whereas Sumba and Timor in the outer arc are primarily of sandstones, mudstones with some igneous intrusions overlain by relatively recent limestone.

The highest mountain in the Lesser Sundas is the impressive Rinjani volcano at 3,726 metres (12,224 feet) on Lombok, followed by one of the limestone peaks on Timor which reaches 2,960 metres (9,711 feet). East of Lombok is the island of Sumbawa on which the highest peak is Tambora at 2,821 metres (9,255 feet). The dramatic eruption in 1883 of Krakatau between Sumatra and Java holds the world record for the greatest explosion, but the world's greatest volcanic eruption in historic times was of Mount Tambora in 1815. The colossal quantity of rock, lava and ash that was thrown out left the mountain 1,250 metres (4,100 feet) lower and with a crater 11 kilometres (nearly 7 miles) across. Some 90,000 people died as a result of the explosion, and the ash falling thickly in the region had serious effects on the vegetation which in turn resulted in the deaths of a great many animals.

The Lesser Sundas is divided into three provinces: West Lesser Sundas (20,200 square kilometres or 7,800 square miles) comprising the islands from Lombok to Sumbawa; East Lesser Sundas (47,900 square kilometres or 18,494 square miles) from Flores and Sumba to West Timor and Alor; and East Timor (15,400 square kilometres or 5,946 square miles). The major towns are Mataram on Lombok, the capital of West Lesser Sundas, and Kupang on the southern tip of Timor which serves as the capital of East Lesser Sundas. The youngest province is East Timor, capital Dili, which occupies the eastern half of Timor and was incorporated within the Republic after the collapse of the former colonial Portuguese Administration of 1976. The Lesser Sundas has a population of about 7.5 million comprising two distinct races: Malays who predominate in the west and Melanesians who predominate in the east. There are also immigrants of Buginese (South Sulawesi) and Chinese origin.

This is the driest part of Indonesia, receiving an average annual rainfall of 1,349 millimetres (53 inches) distributed unevenly through the year. What little rain there is falls from December to March but, even so, this is the most seasonal part of the country. The wettest parts are in some of the mountainous areas of the larger islands, but even these can only really be regarded as moist.

Given its strikingly different climate, it is hardly surprising that the vegetation of the Lesser Sundas is largely different from that in the rest of the archipelago. Thus the natural vegetation would be monsoon forests and savannah woodlands with moist evergreen forest in the mountains. However, little remains of the drier forest types. The practice of starting fires in the dry season to clear the understorey and encourage new growth to provide forage for domestic animals has resulted in a huge loss of natural forest. Inevitably, young trees and seedlings are killed by the fires and, in time, with no new trees to replace the ageing forest, a grassland is formed. Most of the remaining forest is confined to the steepest slopes and the tops of mountains. Floristically the Lesser Sundas is the least rich part of Indonesia with only 12 per cent endemism, most of these species occurring on Lombok and Timor.

The land of the Lesser Sundas is agriculturally productive in the west and generally infertile in the east, where the harsh environment has produced some interesting adaptions. On the small island of Roti, south of Timor, the people have developed a diet and economy based entirely on the sugar palm. Elsewhere the major food crops are rice, maize, soya bean and cassava, while the major cash crops are coconuts and coffee. Fishing is increasingly important but much of the work and processing is on board ship and there is little direct impact on the land. Timber is of only minor importance.

The most famous wild area in the Lesser Sundas is Komodo National Park, comprising Komodo Island and neighbouring islands. This is now an important tourist destination and attracts or diverts many visitors *en route* to Bali. Many visitors also climb Mount Rinjani on Lombok, the neighbouring island to Bali and a 'new' travellers' discovery. Sumbawa has extensive areas of forest on its western hills and mountains and is interesting to visit, but access is far from easy.

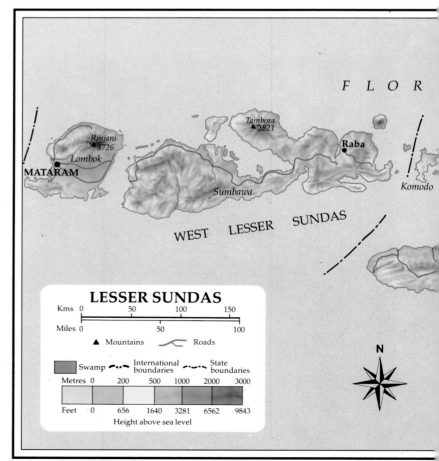

LESSER SUNDAS

Kms 0   50   100   150

Miles 0   50   100

▲ Mountains   Roads

Swamp   International boundaries   State boundaries

Metres 0   200   500   1000   2000   3000

Feet 0   656   1640   3281   6562   9843

Height above sea level

Rain forest on the western slopes of Mount Rinjani (3,726 metres or 12,224 feet), the highest mountain in the Lesser Sundas and the dominating feature on the island of Lombok.

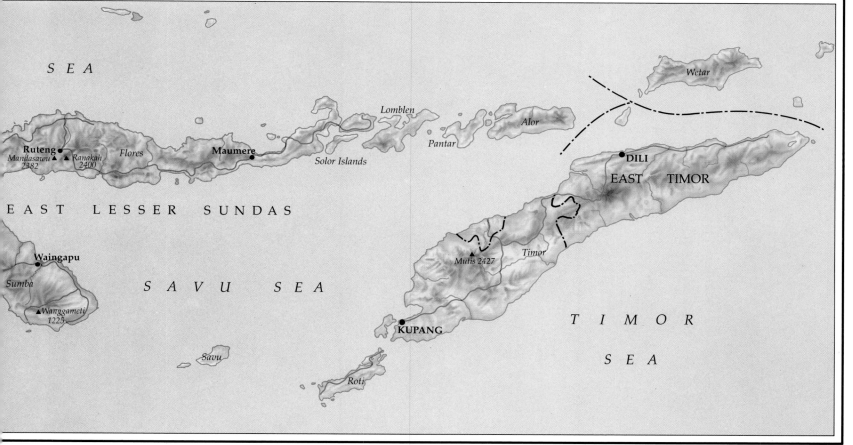

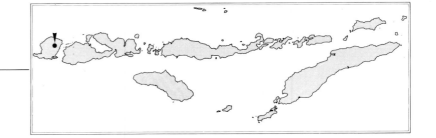

Mount Rinjani National Park

Mount Rinjani, 3,726 metres (12,224 feet) dominates the landscape of Lombok, the island immediately to the east of Bali and accessible from there by air and ferry. The park, 400 square kilometres (155 square miles) in extent, is centred on this mountain, the second highest in Indonesia outside Irian Jaya, but the surrounding 760 square kilometres (over 290 square miles) of forests and scrub on the slopes of adjacent mountains and ridges are also legally protected. Rinjani itself has an impressive 10-kilometre (6-mile) wide crater at the bottom of which is the shallow Segara Anak lake, the water from which flows out in a spectacular waterfall to the north-east and along steep gorges to the north coast. This lake can now be reached by road, and it can only be hoped that this increased accessibility will not destroy the very values which make this wild mountain such a dramatic national park. Man has had considerable impact on the vegetation, but there are still some patches of primary semi-evergreen rain forest to the south and west of Mount Rinjani. This is the most easterly rain forest in the Lesser Sundas. To the east, the forest is of a dry monsoon type with acacia trees more reminiscent of African plains than of Indonesia. Among the 109 bird species recorded are Australian species such as the Sulphur-crested Cockatoo and several honeyeaters, but the Australian influence is not present among the mammals. As Lombok continues to become more popular as a holiday destination, so the number of visitors to the park will increase. Accommodation and interpretation centres are currently lacking but plans are afoot to improve the facilities. It is also hoped that all the remaining forest will be incorporated within the park while measures are taken to ensure that the local people's needs for firewood and other forest products are met.

RIGHT Segara Anak lake backed by the towering walls of the crater.

BELOW Beard 'moss' (*Usnea*) draping over the branches of *Casuarina junghuhniana* trees. It is in fact a lichen, a composite plant comprising a fungus matrix with algal cells within it. *Usnea* has been used in medicines since classical Greek times, and in Indonesia its astringent properties are used to treat intestinal problems.

ABOVE Two outlandishly patterned figs of the creeping species *Ficus aurantiaca*.

RIGHT Hot, sulphurous spring by Segara Anak lake. Note the sparseness of the *Casuarina* cover, resulting from frequent man-induced fires.

OPPOSITE PAGE Segara Anak, the lake in the crater of Mount Rinjani.

Komodo National Park

Komodo Island, in the Lesser Sundas, is one of the most famous wildlife areas in Indonesia, thanks to the enormous Komodo Dragon, the world's heaviest lizard. The national park has a total area of 750 square kilometres (290 square miles) encompassing more or less the entire range of the lizard on the small islands of Komodo, Padar, Rinca and Gili Motong as well as part of mainland Flores and surrounding marine areas. Much of the park is dry and rugged, particularly around the Komodo peak of Mount Toda Klea, and the irregular coastline has many rocky headlands and sandy bays, some framed by vertical cliffs. The seas around the islands are said to be very productive because of the upwelling of deep waters rich in nutrients and the high oxygenation of the water caused by racing tidal currents. Many of the reefs have suffered at the hand of man, but the best remaining ones are on the north-east coast of Komodo and the south-west coast of Rinca and Padar. The most striking tree here is the Lontar Palm which stands in grasslands. Some monsoon forest can be seen where it has escaped fire at the bottom of hills and in valley bottoms and it contains a few endemic species. All the large conspicuous mammals have been introduced by man, but indigenous frogs, snakes and lizards live in the montane forests. The birds show a distinct Australian influence, with the Sulphur-crested Cockatoo, Friar Bird and Scrubfowl being found in the forest fragments. The most serious problem facing the park is the maintenance of the lizards' main prey species of deer and wild boar which are threatened by attacks from feral dogs and from poaching. Fires set by man are a serious problem on the west of Komodo because soil washes off the burnt slopes causing siltation and the death of coral in the sea below. Fears have also been expressed that the baiting of the lizards near the park headquarters may be causing the animals to lose their healthy suspicion of man, with possibly serious consequences.

ABOVE AND OPPOSITE PAGE Landscapes of Rinca Island, which lies between Komodo and the Flores mainland, all of which are part of the Komodo National Park. Note that forest is restricted to valley bottoms and the coast, the two regions where fire is least frequent. More than half of the park is either open grassland or savannah woodland. These islands still have some good coral reefs off their shores.

LEFT *Rhizophora* mangrove off a small island adjacent to Rinca Island. The tangled stilt roots provide a refuge for small fishes.

BELOW The Lontar Palm (*Borassus flabellifera*) is a very common tree on both Komodo and Rinca Islands (as elsewhere in the Lesser Sundas), as the mature trees are resistant to the frequent fires. Note that there are few small palm trees, indicating that in due course the landscape will be far less interesting unless the fires are controlled.

LEFT AND ABOVE Komodo Dragons (*Varanus komodoensis*) are both predators and scavengers, taking deer, pig, young buffalo and monkeys as the major prey species. Although some Komodo Dragons are becoming used to people, most of the some 5,000 individuals in the park are very wary and will run away quickly from possible danger. If cornered, however, they can become very aggressive.

ABOVE These cantering feral horses on Rinca Island are descended from stock introduced by the Sultan of Bima last century for breeding purposes. There are about 300 horses on the island and they appear in good condition, thriving on the grassy hills.

RIGHT *Vanda limbata*, a lowland forest species of orchid. Prior to this photograph being taken it was known only from Java and the Philippines.

Focus on Irian Jaya

Irian Jaya comprises the western half of the large island of New Guinea situated just north of the northern tip of Australia. The eastern half of the island is the country of Papua New Guinea. The islands of Biak, Waigeo, Misool and Salawati also fall within the administration of the province. About 45 per cent of the province is hilly or mountainous and about ten major peaks reach over 4,000 metres (13,120 feet), the tallest of which is Puncak Jaya or Carstenz Peak at 5,039 metres (16,532 feet) from which flows a glacier. Vast stretches of the province remain unexplored, especially in the uninhabited lowlands. Two of the rivers are over 500 kilometres (300 miles) long (and rival the large rivers in Kalimantan as the longest rivers in Indonesia), the Mamberamo-Tariku-Taritatu which flows to the north coast, and the Digul which flows to the south coast close to the large swampy island of Dolak. Irian Jaya stretches 2,500 kilometres (1,550 miles) from east to west and is Indonesia's largest province, covering 414,800 square kilometres (160,150 square miles), three times the combined area of Java and Bali, although there are expectations that it will not be too much longer before this will be divided into four or five separate provinces. There are ambitious plans for a road network but for a good while yet the chief long-distance transport will be small aircraft for which literally hundreds of landing strips exist throughout the island, often built by missionaries.

The population of Irian Jaya is about 1.5 million. In marked contrast to the other Indonesian islands, where most of the people live around the coasts, about 40 per cent of the population lives in the cool central highlands around Wamena and the Paniai Lakes. The indigenous people of Irian Jaya are Melanesians with very dark skins and curly hair. They generally have a root crop subsistence agriculture based on taro and sweet potato. People in the lowlands and swamps, however, obtain their starch from the sago palm which gives an extremely generous yield for remarkably little effort. Feral and domesticated pigs are also important in the local economy. Until fairly recently, many of the people lived within a simple Stone Age culture wearing little clothing and decorating their bodies with tattoos, paintings, shells, pig tusks, feathers and skins. There is a plethora of languages in the province, perhaps some 250 in all, each representing a tribal group which mixes little with the others. Some of the more remote groups still have virtually no contact with the outer world. Indonesians from elsewhere in the country have come to the province either independently seeking new opportunities in a frontier setting, as government-sponsored migrants from Java and Bali, or as government employees. The missionary influence has been stronger here than anywhere else in the country and missionaries have opened up the interior, allowing education and other government facilities to reach the people.

Most of Irian Jaya is wet all year round, but in the south the land is in the Australian rain shadow and is much drier and more seasonal, with the result that the vegetation develops into savannah woodland which is very susceptible to fire. The average annual rainfall is 3,185 millimetres (125 inches), but

The Sudirman Mountains form the backbone of Irian Jaya. Rising up to nearly 5,000 metres (16,400 feet), they provide habitats for many unique species.

there are weather stations in the mountains that receive nearly 6,000 millimetres (240 inches), and stations around the coast receiving little more than 1,500 millimetres (60 inches). At higher altitudes the night-time temperatures fall below zero.

The greater part of Irian Jaya is still covered by some form of forest and little of this has yet been logged. Much of the remainder is bush, scrub and grasslands which are concentrated in the Baliem Valley in the central highlands and in the area around Tanahmerah near the southern border with Papua New Guinea.

High on the southern slopes of the highest mountains in the main range is one of the world's most spectacular mines, the 'Ore Mountain' mine, which sits on the world's largest and purest lodes of copper and from the spoils of which are retrieved large quantities of gold and silver. The ore is taken away by the world's longest single-span cablecar track to a massive mill from which runs the world's longest slurry pipeline. Over 25 million tons of copper have been extracted here, and as much remains to be removed. There are both fears and hopes (depending on one's point of view) that further rich ore deposits will be found to the east in the proposed Mount Lorentz National Park.

As the most remote and uncontestably the wildest part of Indonesia, Irian Jaya offers wilderness experiences beyond anything else in the region. The grandest area is the enormous Mount Lorentz Reserve which stretches from the glacier-bound mountains of the central range all the way down to the coast, encompassing a wide range of ecological communities. The complementary reserve in the north is Mamberamo-Foja which incorporates the virtually uninhabited Mamberamo river and freshwater swamp system, and the untouched Foja mountains where the animals are said to be uniquely tame. To the west is the magnificent Arfak Reserve with its many endemic species, to the south-east the totally different Wasur National Park with its monsoon forest and wet grasslands which attract a vast range of waterbirds.

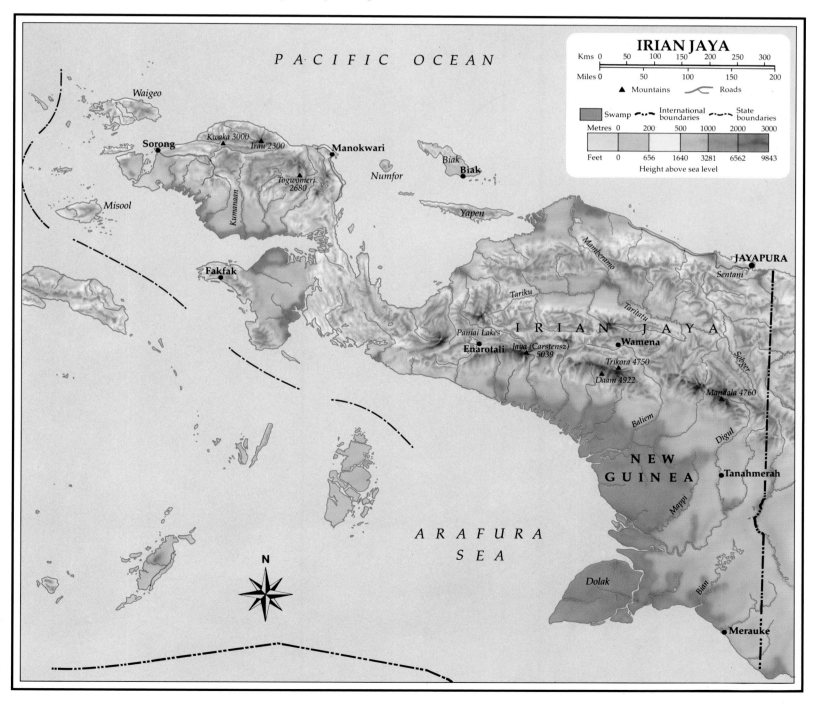

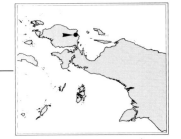

Arfak Reserve

The Arfak Reserve (450 square kilometres or 174 square miles) lies in the north-east corner of the Bird's Head Peninsula of Irian Jaya just 25 kilometres (15 miles) away from the regional capital of Manokwari. It extends from close to the coast up to the highest peak in the area, Mount Humibou at 2,802 metres (9,193 feet). The vegetation comprises lowland, hill and montane rain forests, the latter being rich in oaks, chestnuts and southern beeches. Some of the lowland and hill forest has been selectively logged but it has retained much of its diversity. The reserve protects large numbers of endemic, rare and commercially important species. For example, surveys suggest that about 110 species of mammal may be found here, including thirty species of marsupial – twenty-one of them, including some tree kangaroos, forest wallabies, bandicoots, possums and cuscus, endemic to New Guinea. Of the twenty-seven species of rodent present, seventeen are endemic to New Guinea. About half of the 320 bird species recorded for the reserve are endemic to New Guinea, and some of them are not known outside the Arfak region. In addition, one of the most spectacular insects present is the Rothschild's Birdwing which is also only known from the Arfak area. A singularly successful conservation project has brought the indigenous communities into the management of the reserve and some of its boundaries; and while local institutions remain in force, the positive effects should continue. Even so, not all the reserve is covered in this way and demarcation of boundaries without local consultation has led to serious problems.

Approaching storm over Doreri Bay outside Manokwari, gateway to the reserve.

ABOVE Lower montane forest with abundant tree ferns at about 2,000 metres (6,560 feet).

RIGHT Pristine lower montane forest between Manokwari and Minyambou.

BELOW Waterfall in the Minyambou region of the reserve.

ABOVE The Spotted Cuscus (*Spilocuscus maculatus*) is the most common cuscus in Irian Jaya.

ABOVE RIGHT The Grizzled Tree-kangaroo (*Dendrolagus inustus*) is found in lowland forest up to 1,400 metres (4,590 feet) along the north of the island and the western peninsulas. It is less well adapted to life in the trees than the other tree-kangaroos, and as a result is more easily caught. In the more densely populated areas its populations may be threatened.

BELOW AND BELOW LEFT The Black Tree-kangaroo (*Dendrolagus ursinus*) is an inhabitant of lowland forest and appears to be confined to the Bird's Head Peninsula and the Bomberai Peninsula below it. Despite its name, it sometimes forages on the forest floor, and probably includes plant roots and small animals in its diet of tree leaves and fruit. The fur at the side of the face is white on some animals and rufous on others.

ABOVE The Lesser Bird of Paradise (*Paradisaea minor*) is an inhabitant of the forest-edge, secondary growth both in lowland and lower montane forests. The flamboyant males form groups which display wildly to the rather drab females. After mating the male has no further involvement with the female or the chicks.

BELOW The Amethystine Python (*Morelia amethistina*) is found in northern Australia, New Guinea and the Moluccas.

ABOVE The Yellow-faced Myna (*Mino dumontii*) is a distinctive bird with white patches under the wings, similar to many other Myna species. Although primarily a fruit eater, it will also take advantage of the occasional swarms of flying termites. Mynas are generally seen in pairs or small groups at the forest edge or even in large garden trees.

LEFT Exceptional two-toned *Rhododendron zoelleri* at about 2,000 metres (6,560 feet). This species occurs from Seram to New Guinea and can grow epiphytically as well as in the soil like this specimen. It is found both in forests and in grassland.

BELOW LEFT *Drimys piperata* is a widespread shrub or small tree occurring from Borneo east to Australia and the Solomons in both primary and secondary habitats from 800 metres (2,625 feet) to over 4,000 metres (13,120 feet).

BELOW An unidentified species of *Rhododendron*.

Mount Lorentz Reserve

Spectacular Mount Lorentz Reserve in central Irian Jaya is the largest conservation area in Indonesia, covering some 21,500 square kilometres (8,300 square miles). Its major feature is the exceptional range of habitats represented, from the glacier-covered Puncak Jaya or Carstenz Peak, Indonesia's highest mountain, down through formidably steep and rugged valleys and many types of forest to the freshwater swamps, plains and torrid mangrove swamps on the south coast. This is one of only three places in the world where glaciers can be found at equatorial latitudes (the others being East Africa and the Andes). The high mountains are of deeply weathered limestone, to the north of which are extensive alpine plateaux dominated by tree ferns. In fact there are thirty-four vegetation types in the reserve, encompassing all the major environments recognized in Irian Jaya. Over 120 species of mammal have been recorded, equivalent to 80 per cent of the entire Irian Jaya mammal fauna. As one might expect, the bird fauna is also very rich, with 411 species recorded, including at least twenty species endemic to Irian Jaya such as the Snow Mountain Quail, Orange-cheeked Honeyeater and long-tailed Paradigalla Bird of Paradise. Doubt looms over the future integrity of this internationally important reserve, however, because sitting shoulder to shoulder with it to the west is the massive open-cast Freeport copper mine. Although there is a great deal more ore to extract from the lode currently being exploited, efforts are being made to excise some land from the Lorentz Reserve to allow for mining in the future.

The glacier-capped Puncak Jaya (5,039 metres or 16,532 feet) in the Sudirman Mountains at the western end of the reserve.

ABOVE Cloudy montane forest at about 2,500 metres (8,220 feet) festooned with moss and other epiphytes.

ABOVE RIGHT Spider nurseries decorating rhododendron bushes at 2,300 metres (7,545 feet) in the reserve. If the webs were disturbed, myriads of infant spiders would swarm out of them.

RIGHT An undescribed species of tree-kangaroo, *Dendrolagus*, from near Tembagapura in the Sudirman Mountains.

ABOVE Meandering river with an ox-bow lake in the south of the reserve.

RIGHT Aerial view of a remote valley in the alpine zone.

TOP AND ABOVE Tiny orchids 4 centimetres (1 ½ inches) tall. *Dendrobium vannouhuysii* is known only from Irian Jaya and *Dendrobium subclausum* is a very variable orchid found through the New Guinea mountains. Both species are pollinated by small birds, such as sunbirds, and can be found growing either epiphytically or in the ground.

ABOVE Pitcher plant, *Nepenthes maxima.*

BELOW The fruits of a species of wild ginger (*Riedelia*) at 2,000 metres (6,560 feet).

A beautiful claret-coloured *Rhododendron.*

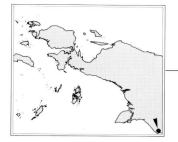

Wasur National Park

The Wasur National Park (4,260 square kilometres or 1,645 square miles) lies in the far south-east of Irian Jaya immediately adjacent to the Tonda Reserve in Papua New Guinea. It is easily accessible from Merauke along the Trans-Irian Highway which, in fact, bisects the park from east to west. Wasur is a largely flat expanse of wetland and coastal plain comprising permanent and seasonal lakes and marshes criss-crossed by numerous rivers. The park stretches from the mangrove-fringed coast, through beech forests, impenetrable bamboo forests, and swamps, to the open savannah woodland and mixed monsoon forest character-istic of the seasonally flooded plains. Fire often sweeps through the inland forests and savannahs during the dry season. Nearly 400 species of birds have been recorded, many of them migratory waders, ducks, cranes, storks, ibis, spoonbills and other water birds. Also seen here are Agile Wallabies, and the endemic Salvador's Monitor Lizard, the longest lizard in the world. Rather more elusive are the endemic Plateless Turtle, the Frilled Lizard, and a wide variety of frogs and toads. Within the forests a patient observer may also be able to see birds of paradise, riflebirds, crowned pigeons and a host of other birds. There are moves ahead to begin developing the park, with the help of the World Wide Fund for Nature, paying particular attention to the socio-economic welfare of the local people, and to preventing further damage from logging, burning of grass-lands and the grazing of domestic and introduced animals. It is expected that increasing numbers of domestic and foreign tourists will visit the reserve in the future, and the inevitable problems that this will bring need to be planned for now.

This savannah of *Imperata* grassland with *Livistona* palms and peeling Paperbark trees (*Melaleuca leucodendron*) gives the impression of being a managed parkland. It is, in fact, the result of frequent fires in a formerly more closed woodland.

ABOVE Group of primitive cycads (*Cycas* sp.). While they look like palms or ferns, they are related to an extinct group of plants known as seed-ferns which flourished about 200 million years ago.

ABOVE LEFT Dense forest of Paperbark trees (*Melaleuca leucodendron*).

OPPOSITE PAGE Sunrise over the Blater swamps.

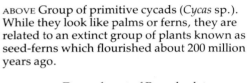

FAR LEFT This flower-like structure is the empty fruit of a *Dillenia* tree.

LEFT This photograph of the orchid *Dendrobium canaliculatum* on a Paperbark tree is the first record for the species in Irian Jaya.

ABOVE Large-buttressed fig trees (*Ficus*) are cut and the sticky white sap that flows from the wounds is used by local people to trap birds.

ABOVE The buds and flowers of a wild rose-apple (*Syzygium*) grow directly from the trunk; this allows birds and bats better access to the flowers and fruit for pollination and dispersal.

ABOVE A 3-metre (10-foot) termite mound in the dry savannah. These impressive constructions by such small creatures are all the more remarkable for the depth to which they continue under the ground.

ABOVE Part of a large mixed flock of waterbirds including Pied Heron (*Egretta picata*), Straw-necked Ibis (*Threskiornis spinicollis*), Glossy Ibis (*Plegadis falcinellus*), Great Egret (*Egretta alba*) and Black-winged Stilt (*Himantopus himantopus*).

BELOW An immature White-bellied Sea-eagle (*Haliaeetus leucogaster*) surveys the plains for prey.

ABOVE Brolga Cranes (*Grus rubicundus*) dancing prior to breeding in the relatively shallow swamps. The nesting season occurs during the wet season from December to May. These large birds apparently avoid deeply flooded, muddy or densely vegetated areas.

ABOVE Male Red-sided Eclectus Parrot (*Eclectus roratus*) a conspicuous and noisy bird of lowland forest and savannah. The sexes look very different: the female Eclectus is bright red. They are strong fliers and can sometimes be seen high above the forest canopy, using a few wing-beats followed by a glide.

ABOVE RIGHT Masked Lapwing (*Vanellus miles*) at its nest; these pugnacious birds generally mob intruders who come too close to their eggs. Outside the breeding season they form large aggregations in the swampy lowlands. They range from New Guinea to New Zealand, although they have only established themselves in New Zealand in recent years.

BELOW A strutting Australian Bustard (*Ardeotis australis*). These birds are common in the short grassland plains. Males display around sunrise and their loud booming calls can be heard from far off. Their size, calls and approachability have made them easy targets for hunters.

FAR RIGHT The Southern or Two-wattled Cassowary (*Casuarius casuarius*) is the largest of the three cassowary species in New Guinea, weighing up to 60 kilograms (130 pounds). It is quite common within the park, where it feeds primarily on fallen fruit. The swampy country is no deterrent to this bird because it swims readily. It is also unusual in that the male alone cares for the eggs and chicks.

ABOVE LEFT Large monitor lizard (*Varanus salvator*) emerging from its tree hole. The long, but small-bodied, endemic Salvadori's Monitor (*Varanus salvadorii*) also occurs in the park but is rarely seen.

ABOVE The Short-beaked Echidna (*Tachyglossus aculeatus*) is found in southern and eastern New Guinea. It has no teeth and uses its long sticky tongue to feed on the eggs and pupae of ants and termites, breaking into their nests with its large claws.

LEFT A widespread and striking tiger moth (*Euchromia creusa*).

BELOW LEFT The common dragonfly, *Neurothemis terminata*, at rest.

BELOW An unspotted race of the Spotted Cuscus (*Spilocuscus maculatus*). This variable species is very widespread being found throughout New Guinea, west to Seram and Buru, and south to the northern tip of Australia. It is entirely arboreal and nocturnal, and eats leaves and fruit.

ABOVE AND RIGHT The Agile Wallaby (*Macropus agilis*) is found in the monsoonal plains and savannahs of southern New Guinea, and some 30,000 may occur in the park. They are hunted (illegally), their persecutors burning the grass during the dry season and waiting for the wallabies when they return to graze on the new growth. This species also eats grass roots, *Eucalyptus* leaves and figs. Breeding occurs year round and the young remain in the pouch for up to eight months and continue suckling until a year old.

Major Conservation Areas

These areas are shown on the map on pages 14 and 15 and are listed here in either approximately north to south or west to east order. The size of each area is given but different source documents give different figures. Not all these areas have yet been gazetted and may currently be being destroyed.

SUMATRA

Mount Leuser National Park, 9,500 square kilometres (3,668 square miles), one of the great forest conservation areas in South-east Asia.

Singkil Barat, 851 square kilometres (329 square miles), an obvious extension to Mount Leuser. NOT YET GAZETTED.

Dolok Sembilin, 339 square kilometres (131 square miles) another obvious extension to Mount Leuser. NOT YET GAZETTED.

Taitai Batti, 565 square kilometres (218 square miles), the only reserve protecting the unique wildlife of the Mentawai Islands.

Taitai Batti southern extension. NOT YET GAZETTED.

Kerumutan Baru, 1,200 square kilometres (463 square miles), to extend the existing Kerumutan lowland and swamp forest reserve. NOT YET GAZETTED.

Seberida, 1,200 square kilometres (463 square miles), an area of hills separated from the main range of mountains. NOT YET GAZETTED.

Bukit Besar, 300 square kilometres (116 square miles), a logical extension of Seberida. NOT YET GAZETTED.

Datuk Point and Bakung Island, 550 square kilometres (212 square miles), a good area of wetland coast and peat swamp forest. NOT YET GAZETTED.

Jabung Point, 30 square kilometres (12 square miles), a small area of special interest for wading birds. NOT YET GAZETTED.

Berbak Reserve, 1,900 square kilometres (734 square miles), an internationally important area of peat swamp.

Kerinci-Seblat National Park, 14,846 square kilometres (5,732 square miles), contains western Indonesia's highest mountain and the full range of wildlife.

Kembang Lubok Niur, 1,000 square kilometres (386 square miles), a northern extension to Kerinci-Seblat. NOT YET GAZETTED.

Banyuasin-Musi/Sembilang River, 900 square kilometres (347 square miles), a special area of mangroves and mudflats. NOT YET GAZETTED.

South Barisan National Park, 3,568 square kilometres (1,378 square miles), a major area of mainly montane forest at the tip of the Barisan Mountains.

Way Kambas National Park, 1,300 square kilometres (502 square miles), a very disturbed area of freshwater swamp forest

with much potential if protected effectively.

JAVA

Ujung Kulon National Park, 786 square kilometres (303 square miles), the most famous park in Indonesia because of its population of Javan Rhinoceros and beautiful scenery.

Mount Halimun Reserve, 400 square kilometres (155 square miles), one of the largest areas of forest left in Java and much undervalued.

Mount Gede-Pangrango National Park, 150 square kilometres (58 square miles), a famous mountain park with many interesting species and a spectacular crater.

Thousand Islands Marine Park, 1,100 square kilometres (425 square miles), small coral islands to the north of Jakarta.

Anakan Lagoon, an important wetland area of the lagoon at the mouth of the Citanduy and other rivers on the south coast. NOT YET GAZETTED.

Mount Kawi-Kelud, 776 square kilometres (300 square miles), an impressive volcanic area with good forests. NOT YET GAZETTED.

Bromo-Tengger/Mount Semeru National Park, 580 square kilometres (224 square miles), centred on the famous Bromo Crater with wonderful scenery but limited wildlife interest.

Meru Betiri National Park, 500 square kilometres (193 square miles), a beautiful area of forest with sheer cliffs facing the Indian Ocean; turtle beaches.

Baluran National Park, 250 square kilometres (97 square miles), at the north-east tip of Java with many large mammals but with serious conservation management problems.

Alas Purwo, 620 square kilometres (240 square miles), Java's largest gazetted terrestrial reserve with large populations of Banteng, Peafowl and Leopard, the first of which can be seen from a watch-tower; the southern tip is said to have the world's best surfing.

BALI

Bali Barat National Park, 777 square kilometres (300 square miles), a surprisingly large area of forest (10 per cent of Bali's area) best known for its endangered Bali Myna.

KALIMANTAN

Mount Palung National Park, 300 square kilometres (116 square miles), a complete spectrum of vegetation types from coasts to peaks of 1,160 metres (3,806 feet). Due to the superstitions of the local villagers, the hill forests are quite undisturbed.

Mount Bentuang/Karimun, 6,000 square kilometres (2,317 square miles), a large area of well forested hills along the

Sarawak border which would constitute a transfrontier reserve. NOT YET GAZETTED.

Lake Sentarum, 800 square kilometres (309 square miles), a unique system of shallow interconnecting lakes surrounded by swamp forests on deep peat. NOT YET GAZETTED.

Bukit Baka, 1,000 square kilometres (386 square miles), lush forest adjacent to the existing reserve of Bukit Raya. NOT YET GAZETTED.

Tanjung Puting National Park, 3,050 square kilometres (1,178 square miles), mangrove, peat swamps, freshwater swamps and heath forest, and best known for its research and Orangutan rehabilitation centre.

Tanjung Puting Extension, 800 square kilometres (309 square miles), an important extension to the west of the park. NOT YET GAZETTED.

Bukit Raya, 1,100 square kilometres (425 square miles), excellent range from little disturbed lowland to montane forests on rich soils and volcanic rocks. Bukit Raya is the highest mountain in Kalimantan and has the best developed montane flora.

Bukit Raya Extension, 5,900 square kilometres (2,278 square miles), extension would add a large area of valuable lowland rain forest to the Bukit Raya reserve. NOT YET GAZETTED.

Kayan Mentarang, 16,000 square kilometres (6,178 square miles), a huge area along the Sarawak border and includes a range of habitats from montane to lowland but is predominantly hill forest.

Ulu Sembakung, 5,000 square kilometres (1,930 square miles), hilly forests ranging from lowland to montane in the botanically rich north-east corner of Borneo, and extension of Kayan Mentarang. NOT YET GAZETTED.

Muara Sebuku, 1,100 square kilometres (425 square miles), a large and little disturbed area of mangroves. NOT YET GAZETTED.

Sangkulirang, 2,000 square kilometres (772 square miles), scenically beautiful and botanically unique forested hills of limestone karst including some traditional burial areas of original inhabitants. NOT YET GAZETTED.

Kutai National Park, 2,000 square kilometres (772 square miles), much disturbed by huge fires, logging and other developments, but still with potential.

SULAWESI

Tangkoko Reserve, 87 square kilometres (34 square miles), a small but valuable reserve with very high densities of macaques and other frugivores.

Bunaken Island, marine reserve encompassing the famous reefs of Bunaken itself together with the reefs around the three adjacent islands.

Dumoga-Bone National Park, 3,000 square kilometres (1,158 square miles), a major conservation area with most of Sulawesi's endemic wildlife; a centre for research.

Marisa, 940 square kilometres (363 square miles), a complex of three reserves protecting important forest for maleos, primates and wetlands. NOT YET GAZETTED.

Morowali National Park, 2,000 square kilometres (772 square miles), a unique mosaic of forest on ultrabasic soils, limestone, and alluvial plains.

Lore Lindu National Park, 2,290 square kilometres (884 square miles), major conservation area with most of Sulawesi's endemic wildlife and an interesting lake.

Togian Islands, marine area with a wide range of reef types and large populations of coconut crabs. NOT YET GAZETTED.

Rawa Aopa/Watumohae National Park, 968 square kilometres (374 square miles), a large swamp, forested hills, and coastal forest with potential for boosting tourism in South-east Sulawesi.

Matano-Mahalona Lakes, 300 square kilometres (116 square miles), and Towuti Lake, 650 square kilometres (250 square miles), exceptional lakes with many endemic fishes, crabs, snails, etc, but current protection status is inadequate.

Mount Latimojong, 300 square kilometres (116 square miles), little known forested mountains including the highest peak in Sulawesi; good potential for broadening experience of tourists visiting Torajaland. NOT YET GAZETTED.

Sarege, 2 square kilometres (¾ square mile), a small island with rats and no fresh water, but a good seabird breeding colony especially for the Red-footed Booby; also has original beach vegetation. NOT YET GAZETTED.

Kakabia, 2 square kilometres (¾ square mile), small islands with large numbers of nesting seabirds, excellent undisturbed beach vegetation, and no fresh water. NOT YET GAZETTED.

LESSER SUNDAS
Mount Rinjani National Park, 400 square kilometres (155 square miles), the crater and surrounding forests of the highest Indonesian peak outside Sumatra and Irian Jaya.

Komodo National Park, 750 square kilometres (290 square miles), group of islands and part of Flores mainland with coral reefs; home of the Komodo Dragon.

Ruteng, 300 square kilometres (116 square miles), forested hills with exceptional scenery and endemic birds. NOT YET GAZETTED.

Mount Olet Sangenges, 350 square kilometres (135 square miles), lush forest with many endemic species. NOT YET GAZETTED.

Tambora Complex, 1,100 square kilometres (425 square miles), northern peninsula including the site of the world's largest volcanic eruption in 1815, and forests with endangered habitats and endemic wildlife. NOT YET GAZETTED.

Mount Wanggameti, 60 square kilometres

(24 square miles), a small area but the best on Sumba and includes the island's highest mountain. NOT YET GAZETTED.

Mount Mutis/Timau, 250 square kilometres (97 square miles), the only large area of closed forest left in western Timor, particularly valuable for pure wild stands of the plantation tree *Eucalyptus urophylla*. NOT YET GAZETTED.

Ira Lake/Lalore/Yaco, 400 square kilometres (154 square miles), lowland and montane forests, a swamp, large lake, and an island. NOT YET GAZETTED.

MOLUCCAS
Lalobata, 1,890 square kilometres (730 square miles), excellent undisturbed, hilly limestone and ultrabasic forests. NOT YET GAZETTED.

Mount Sibela, 400 square kilometres (154 square miles), the highest mountain in the northern Moluccas, surrounded by thick forests, and conserving the special Bacan fauna and flora. NOT YET GAZETTED.

Mount Kelapat Mada, 1,450 square kilometres (560 square miles), relatively undisturbed apart from hunting trails, the best area on Buru for conserving the island's endemic species and habitats. NOT YET GAZETTED.

Manusela National Park, 1,890 square kilometres (730 square miles), an excellent swathe of forest across from the north to south coasts and including the Moluccas' highest mountain, protecting the many Seram endemic mammals, birds and reptiles.

Wae Bula, 450 square kilometres (174 square miles), in the drier part of Seram with swamps, lowlands and mountains. NOT YET GAZETTED.

Mount Api, 0.8 square kilometre (⅓ square mile), tiny volcanic island with very important colonies of nesting seabirds.

Manuk Island, 1 square kilometre (⅓ square mile), small volcanic island but possibly the most important seabird nesting colony in the region. NOT YET GAZETTED.

Mount Arnau, 450 square kilometres (174 square miles), deciduous monsoon forests on steep volcanics and on limestone with five endemic bird species. NOT YET GAZETTED.

Yamdena, 600 square kilometres (232 square miles), the remaining forested area on the main island of the Tanimbar group to protect the endemic bird species. NOT YET GAZETTED.

Kai Besar, 370 square kilometres (143 square miles), limestone hills on the largest but least disturbed island in the Kais, representing the best area to conserve the islands' endemic species. NOT YET GAZETTED.

Aru Tenggara, 2,000 square kilometres (772 square miles), a complex of marine and coastal habitats such as reefs, seagrass beds, mangroves and beach vegetation, in a little-disturbed part of the archipelago. NOT YET GAZETTED.

Kabroor Island, 690 square kilometres (266 square miles), despite being the largest island in the Aru group, it is the least disturbed and the best area for conservation. NOT YET GAZETTED.

IRIAN JAYA
Lorentz, 21,500 square kilometres (8,301 square miles), largest and most complete reserve in Indonesia, including all habitats from the coast to alpine meadows and glaciers.

Mamberamo-Foja, 14,420 square kilometres (5,568 square miles), wonderful, barely disturbed range of habitats from mangroves and peat swamps, across the Foja Mountains to the Mamberamo plains where the rich wildlife is said to be exceptionally tame. NOT YET GAZETTED.

Rouffaer, 818 square kilometres (316 square miles), western extension to Mamberamo. NOT YET GAZETTED.

Arfak, 450 square kilometres (174 square miles), an isolated and forested mountain block in a dry area of Irian Jaya with some species known only from this area.

North and South Tamrau, 4,415 square kilometres (1,705 square miles), hilly area with some lowlands and an area of ultrabasic rock with stunted forest; contains many rare endemic species. NOT YET GAZETTED.

Wasur-Rawa Biru National Park 4,260 square kilometres (1,645 square miles), excellent area of savannah woodlands, grasslands and wetlands with large numbers of resident and migratory waterbirds.

Cyclops, 225 square kilometres (87 square miles), isolated range of steep hills close to the provincial capital and covering several interesting habitats with high levels of endemism.

Jayawijaya, 8,000 square kilometres (3,089 square miles), large area spanning the central mountains from the virtually uninhabited Taritatu River in the north to the southern swamps. NOT YET GAZETTED.

Bintuni Bay, 2,600 square kilometres (1,004 square miles), extensive mangrove forests.

Sausapor and Jamursba-Mandi Beaches, 85 kilometres (53 miles) of beach used by four species of turtle and including one of the largest Leatherback rookeries in the world. NOT YET GAZETTED.

Cenderawasih Bay National Park, 14,330 square kilometres (5,533 square miles), including 500 kilometres (310 miles) of coast, bird nesting colonies and some of the richest coral reefs in the world.

Bibliography

Listed below is a selection of the books the authors have found useful in learning about Indonesia's natural ecosystems, the background to them, travelling between them, and identifying a range of animals and plants. Some of the books are out of print, but can be found in libraries. Publications for which specialist knowledge of anatomy and morphology are required have, for the most part, not been included.

TRAVEL
Among the best travel guides available from the plethora lined up on bookshop shelves are those produced by Periplus Editions.

Caldwell, I. (1991) *Sumatra: Island of Adventure*. Periplus, Singapore and Berkeley.

Muller, K. (1989) *Irian Jaya: West New Guinea*. Periplus, Singapore and Berkeley.

Muller, K. (1990) *Kalimantan: Indonesian Borneo*. Periplus, Singapore and Berkeley.

Muller, K. (1990) *Maluku: The Moluccas*. Periplus, Singapore and Berkeley.

Muller, K. (1991) *East of Bali: From Lombok to Timor*. Periplus, Singapore and Berkeley.

Muller, K. (1992) *Underwater Indonesia: A Guide to the World's Best Diving*. Periplus, Singapore and Berkeley.

Oey, E. (1989) *Bali: Island of the Gods*. Periplus, Singapore and Berkeley.

Oey, E. (1991) *Java: Garden of the East*. Periplus, Singapore and Berkeley.

Volkman, T. and Caldwell, I. (1990) *Sulawesi (The Celebes)*. Periplus, Singapore and Berkeley.

Wallace, A.R. (1962) *The Malay Archipelago*. Dover, New York.
Every traveller (armchair or otherwise) should delve into this book. Written over 150 years ago, some of the observations are as true now as they were then, and there is a sense of infectious excitement from this contemporary of Darwin.

ECOLOGY
In 1982 a series on the ecology of different regions of Indonesia was begun under a Canadian aid project. The books were written primarily for the Indonesian market of students, lecturers and government planners, but English editions are available too. They contain a wealth of information on ecology (in its broadest sense) and are copiously illustrated.

Whitten, A.J., Damanik, S.J., Anwar, J. and Hisyam, N. (1987) *The Ecology of Sumatra* (2nd edition). Gadjah Mada University Press, Yogyakarta.

Whitten, A.J., Mustafa M., Henderson, G.S. (1987) *The Ecology of Sulawesi*. Gadjah Mada University Press, Yogyakarta.
All the remaining volumes are in preparation: *Kalimantan* written by K.S. MacKinnon *et al.*, *Java and Bali* by A.J. Whitten *et al.*, *Moluccas and Lesser Sundas* by K. Monk *et al.*, *Irian Jaya* by A. Forsyth *et al.*, and *The Seas* by T. Tomascik *et al*. The entire series should be available by late 1993. Although the two earlier titles are not widely available outside Indonesia, they are kept in stock by Heffers Booksellers, Trinity Street, Cambridge, UK, and the Natural History Book Service, Totnes, Devon, UK.

Petocz, R.G. (1989) *Conservation and Development in Irian Jaya*. Brill, Leiden.
In addition, the above book is a masterful compilation of basic information about Irian Jaya, its ecology, peoples, development pressures and possible policy options.

BIRDS
Beehler, B.M., Pratt, T.K. and Zimmerman, D.A. (1986) *Birds of New Guinea*. Princeton University Press, Princeton.
A beautiful book that shows there is a great deal more to New Guinea birds than birds of paradise.

Holmes, D. and Nash, S.V. (1989) *Birds of Java and Bali*. Oxford University Press, Kuala Lumpur.
This is a guide to the commonly seen birds of Java and Bali (in contrast to MacKinnon below). For the casual visitor not visiting particularly wild places this is probably all that will be needed.

Holmes, D. and Nash, S.V. (1990) *Birds of Sumatra and Kalimantan*. Oxford University Press, Kuala Lumpur.
As with the Java and Bali book above, this is not exhaustive but is none the less useful.

King, B., Woodcock, M. and Dickinson, E.C. (1975) *A Field Guide to the Birds of South-East Asia*. Collins, London.
This book covers mainland South-east Asia and not Indonesia. While many of the birds illustrated are found in western Indonesia, many are not.

MacKinnon, J. (1989) *Field Guide to the Birds of Java and Bali*. Gadjah Mada University Press, Yogyakarta.
Indonesia's first indigenous field guide was rightly produced for Java and Bali where most people live and where there is the best hope of raising environmental consciousness about the loss of forest and the species it supports. Almost every species ever recorded on these two islands and their smaller satellites is illustrated. In addition, there are sections on conservation, bird watching etc. Available also in an Indonesian version.

Mason, V. and Jarvis, F. (1989) *Birds of Bali*. Periplus, Singapore and Berkeley.
A very attractive book with life-like paintings of the birds in their habitats. Like Holmes and Nash, it does not try to cover every species, but rather the species that a traveller will see. There are hints on good bird-watching areas.

Smythies, B.E. (1981) *The Birds of Borneo* (3rd edition) (edited by The Earl of Cranbrook). The Sabah Society with the Malayan Nature Society, Kota Kinabalu and Kuala Lumpur.
A classic book, first published in 1960, covering the whole of Borneo but based largely on East Malaysia. As the number of interested ornithologists increases, so a number of species ranges are having to be revised, but this does not detract from the value of the book. Very attractive illustrations and a large map with place names.

MAMMALS – GUIDES
Flannery, T. (1991) *The Mammals of New Guinea*. Australian Museum, Sydney.
A beautiful new book which will open most people's eyes to the large number of mammals on the island – marsupials, rodents and bats.

Medway, Lord (1969) *The Wild Mammals of Malaya*. Oxford University Press, Kuala Lumpur.
A fascinating account of all the mammals in the Malay Peninsula, including notes on diets in captivity, breeding, weights, etc. There is a great deal of similarity between the Malayan mammal fauna and that of Sumatra and Borneo.

Payne, J., Francis, C.M. and Phillipps, K. (1985) *A Field Guide to the Mammals of Borneo*. Sabah Society, Kota Kinabalu and WWF Malaysia, Kuala Lumpur.
A beautifully produced pocket-sized book with exquisite paintings illustrating the species. Don't rely on the paintings to identify the bats or rats, however, since the differences in rendering are not diagnostic. Useful text on the island as well as on the behaviour and ecology of the animals.

MAMMALS – OTHER
MacKinnon, J. (1974) *In Search of the Red Ape*. Collins, London.
The popular account of the first fundamental study of orangutans in both Sabah and in North Sumatra.

MacKinnon, J. (1978) *The Ape Within Us*. Collins, London.
A fascinating view of apes in the wild, such as siamang, gibbons and orangutans, in the context of human evolution.

Whitten, A.J. (1982) *The Gibbons of Siberut*. Dent, London.
An account of the life of gibbons and forest ecology woven into

the story of our life on the remote and fascinating island of Siberut off the west coast of Sumatra.

REPTILES
Lim, F.L.K. and Lee, M.T.M. (1989) *Fascinating Snakes of Southeast Asia*. Tropical Press, Kuala Lumpur.
The most colourful and attractive of a small number of books available. There are seventy-five species illustrated with colour plates of living specimens. This should cover most of the species seen, at least in western Indonesia.

FROGS AND TOADS
Menzies, J.I. (1975) *Handbook of Common New Guinea Frogs*. Wau Ecology Institute Handbook No. 1
With 182 species known from the island of New Guinea (and new ones being discovered) all that can be expected is something like this: a simple guide with colour photographs of those species most likely to be encountered.

FISHES
Allen, G.A. and Cross, N.J. (1982) *Rainbow Fishes of Papua New Guinea and Australia*. T.F.H. Publications, New Jersey.
Gerry Allen and others are discovering new species of rainbow fishes (Melanotaeniidae) all the time, it seems, but this covers most of the species of a group which are becoming very popular aquarium fishes.
Carcasson, R.H. (1977) *A Field Guide to the Reef Fishes of the Indian and West Pacific Oceans*. Collins, London.
Regrettably out of print, but the market is surely not yet satiated as interest in corals increases. Some 1,800 species are described and about 400 representative species are illustrated in colour. The only book covering Indonesia's coral reef fish yet published.
Gloerfelt-Tarp, T. and Kailola, P.J. (1988) *Trawled Fishes of Southern Indonesia and Northwestern Australia*. Australian Development Assistance Bureau, Canberra.
A well-illustrated account of about 1,100 species of marine fish, of great interest if taken round a fish market.
Kottelat M., Whitten, A.J., Kartikasari, S.N. and Wirjoatmodjo, S. (1992) *Freshwater Fishes of Borneo, Sumatra, Java and Sulawesi*, Periplus, Singapore.
This is the only accessible means of identifying all the freshwater and brackish-water fishes in western Indonesia. Almost all of the nearly 1,000 species are illustrated, and most of those are in colour. Useful introductory sections on conservation, ecology, behaviour, pollution and collecting are provided.
Schroeder, R.E. (1980) *Philippine Shore Fishes of the Western Sulu Sea*. NMPC, Metro Manila.
A nice book with colour photographs of most of the more commonly observed coastal species.

BUTTERFLIES AND MOTHS
d'Abrera, B. (1977) *Birdwing Butterflies of the Australian Region*. Lansdowne, Melbourne.
The birdwing butterflies have their centre of diversity in western Irian Jaya, and they have been sought by collectors ever since they were discovered. This book illustrates these often enormous butterflies in all their beauty.
Barlow, H. (1982) *An Introduction to the Moths of South East Asia*. Malayan Nature Society, Kuala Lumpur.
This may be only an introduction, but it serves to identify most of the larger moths a non-specialist is likely to encounter. Simply identifying a moth to a genus is generally very satisfying.
Corbett, A.S. and Pendlebury, H.M. (1978) *The Butterflies of the Malay Peninsula* (3rd edition) (revised by J.N. Eliot). Malayan Nature Society, Kuala Lumpur.
As with the other books on Malayan species, this book will be most useful in Sumatra, and of general use on Java and Borneo. Many of the species illustrated are distributed far wider.
Diehl, E.W. (1980) *Heterocera Sumatrana. Band 1: Sphingidae*. Classey, London.

This covers just the large and dramatic hawkmoths of Sumatra, a group which is surprisingly often encountered.

OTHER TERRESTRIAL INSECTS AND OTHER INVERTEBRATES
Kalshoven, L.G.E. (1981) *Pests of Crops in Indonesia* (revised by P.A. van der Laan). Ichtiar Baru – Van Hoeve, Jakarta.
This book is surprisingly useful for identifying creatures seen flying, creeping, hopping or crawling. It covers everything from worms to monkeys, and is illustrated by line drawings, black-and-white photographs, and colour plates of paintings.

CORAL REEFS
Ditlev, H. (1980) *A Field Guide to the Reef-Building Corals of the Indo-Pacific*. Backhuys, Rotterdam.
Identifying corals is not easy, but the 120 colour photographs of living specimens and the 270 black-and-white photos of dead specimens make it as easy as possible, aided by text descriptions, keys and figures.
Guille, A., Laboute, P. and Menou, J.-L. (1986) *Handbook of the Seastars, Sea Urchins and Related Echinoderms of New Caledonia Lagoon*. ORSTOM, Paris.
Although New Caledonia is a fair way from Indonesia, most of the shallow-sea echinoderms have wide distributions. Some 240 species are illustrated with unexpectedly beautiful colour photographs of living animals. The introductory text and keys are in French and English, but the species descriptions are in French (though an elementary grasp of the language is sufficient to understand most of it).
Henrey, L. (1982) *Coral Reefs of Malaysia and Singapore*. Longman, Kuala Lumpur.
A straightforward introduction to coral reefs and their inhabitants.
Ming, C.L. (1988) *A Guide to the Coral Reef Life of Singapore*. Singapore Science Centre, Singapore.
Ng, L.W.H. and Ng, P.K.L. (1988) *A Guide to Seashore Life*. Singapore Science Centre, Singapore.
Two of a series of small books on different aspects of Singapore's natural history, of considerable relevance to western Indonesia. They cover both the organisms and the general ecology of reefs.

MARINE MOLLUSCS
Dharma, B. (1988) *Siput dan Kerang Indonesia (Indonesian Shells)*. Sarana Graha, Jakarta.
An excellent colour guide to Indonesia's many and varied, colourful and often bizarrely shaped shells. The text is in Indonesian but the species notes are in English too.
Roberts, D., Soemodihardjo, S. and Kastoro, W. (1982) *Shallow Water Marine Molluscs of North West Java*. National Oceanographic Institute, Jakarta.
A useful guide for those fascinated by mangroves and mudflats. Few of the species are particularly colourful, so little is lost by having only black-and-white photographs.

VEGETATION
Whitmore, T.C. (1984) *Tropical Rain Forests of the Far East*. Clarendon, Oxford.
Whitmore, T.C. (1990) *Introduction to Tropical Rain Forests*. Clarendon, Oxford.
The two books on tropical rain forests for those with an interest in Indonesia. Some of the 1984 book may be a bit heavy for the casual reader, but it is extremely sound. Some of the facts may come as a surprise for those whose primary source of information is Friends of the Earth and other campaigning groups.

PLANTS IN VILLAGES AND TOWNS
Corner, E.J.H. (1987) *The Wayside Trees of Malaya*. Malayan Nature Society, Kuala Lumpur.
A classic book now reprinted after years of being unavailable. Incredibly interesting text about the trees one sees in towns,

villages and in rural areas. Good keys.

Eiseman, F. and Eiseman, M. (1988) *Flowers of Bali*. Periplus, Singapore.
A pretty book about the fifty or so flowers one sees in and around Balinese hotels and village compounds.

Lötschert, W. and Beese, G. (1983) *Collins Guide to Tropical Plants*. Collins, London.
Many of the plants grown in gardens around the tropics are the same and so the somewhat audacious title is in fact reasonable if one looks no further than gardens and agricultural areas. There are 325 plant species described with 274 colour photographs.

Polunin, I. (1987) *Plants and Flowers of Singapore*. Times Editions, Singapore.
Similar to the Eisemans' book above, but covering many more species (163) all of which are commonly planted or grow naturally throughout Indonesia.

PLANTS IN RURAL AND FOREST SETTINGS

Coomber, J. (1991) *The Orchids of Java*. Royal Botanic Gardens, London.
All of Java's rich orchid flora illustrated by colour photographs, with full written descriptions and anecdotal extras.

Soerjani, M., Kostermans, A.J.J.G. and Tjitrosoepomo, G. (1987) *Weeds of Rice in Indonesia*. Balai Pustaka, Jakarta.
Grasses, sedges and little herbs are ubiquitous but often hard to identify. This book has very clear botanical line drawings of 266 species which should bring a great deal of order into the otherwise nameless desert of ricefields. The last part of the book has illustrations and descriptions of the seedlings of 226 species – a veritable labour of love.

Steenis, C.G.G.J. van (1972) *The Mountain Flora of Java*. Brill, Leiden.
A wonderful book, not just because of the large, stunning colour plates, but because of the detailed but very readable information about Java's mountains.

Steenis, C.G.G.J. van (1949–83) *Flora Malesiana* (10 volumes on higher plants, one volume on ferns to date). Rijksherbarium, Leiden and Martinus Nijhoff/Junk, The Hague.

While not strictly covering the specialist literature, we must at least make mention of this monumental series of volumes started in 1949 by the late visionary botanist Dr van Steenis. Families are covered one at a time and it will be a decade or more at current rates before the work is complete.

Whitmore, T.C. (1972) *Tree Flora of Malaya* (4 volumes, final volume edited by F.S. Ng). Longman, Kuala Lumpur.

Whitmore, T.C. (1977) *Palms of Malaya* (revised edition). Oxford University Press, Kuala Lumpur.
Again, although intended for use in Malaya, these last two publications are just as useful in Sumatra, and have great utility elsewhere in western Indonesia. Both include text on the multifarious uses to which man has put the plants, and other interesting snippets. Both have very clear identification keys for which no great botanical knowledge is required. The *Tree Flora* covers all trees of timber size (those reaching a girth of 90 centimetres/35 inches), comprising 3,000 species in 98 families.

PLANTS – FERNS AND LOWER PLANTS

Eddy, A. (1990) *A Handbook to Malesian Mosses* (2 volumes so far). Natural History Museum, London.
A somewhat specialist text for a neglected group.

Johnson, A. (1980) *Mosses of Singapore and Malaysia*. Singapore University Press, Singapore.
Similar to the above but less daunting (partly because of the more limited coverage).

Piggott, A. (1979) *Heinemann Guide to Common Epiphytic Ferns of Malaysia and Singapore*. Heinemann, Kuala Lumpur.
Suddenly epiphytic ferns become 'friendly' with this book which quickly allows one to progress beyond just recognising the bird's-nest and stagshorn ferns.

Teo, L.W. and Wee, Y.C (1983) *Seaweeds of Singapore*. Singapore University Press, Singapore.
Seaweeds are really very interesting. With just line drawings this book enables one to tell the common species apart. Almost all the species have a wide distribution.

Index

Page numbers in bold type refer to illustrations; those preceded by m refer to maps.